CAR BRAKES

A GUIDE TO UPGRADING, REPAIR AND MAINTENANCE

CAR BRAKES

A GUIDE TO UPGRADING, REPAIR AND MAINTENANCE

Jon Lawes

THE CROWOOD PRESS

First published in 2014 by
The Crowood Press Ltd
Ramsbury, Marlborough
Wiltshire SN8 2HR

www.crowood.com

British Library Cataloguing-in-Publication Data
A catalogue record for this book is available from the British Library.

ISBN 978 1 84797 674 1

Disclaimer
Safety is of the utmost importance in every aspect of an automotive
workshop. The practical procedures and the tools and equipment used
in automotive workshops are potentially dangerous. Tools should be
used in strict accordance with the manufacturer's recommended
procedures and current health and safety regulations. The author
and publisher cannot accept responsibility for any accident or
injury caused by following the advice given in this book.

Typeset by Servis Filmsetting Ltd, Stockport, Cheshire
Printed and bound in India by Replika Press Pvt. Ltd.

CONTENTS

ACKNOWLEDGEMENTS

The following, arranged in alphabetical order, have been instrumental in the production of this book:

Ben Short of BS Motorsport for his advice and guidance, specifically with regards to driver training and vehicle dynamics

Continental/ATE for their advice and images

EBC Brakes for their assistance with advice and photographs

Eric Jacobsen for his help, photos and advice

Justin Westley for his help and good humour

John Goldsmith of Goldsmith and Young Ltd for his support and loan of facilities

Mark Richardson for his assistance with photos, friendship and support

Mike Simpson for his friendship and photographic assistance

Simon Crosse for his photographic input

Simon Sexton and Adam Gallaway for their handskills and experience in producing the technical sections

And of course my family for their tolerance.

INTRODUCTION

Modern domestic braking systems are designed to a very high standard, from those fitted to the most basic shopping hatchback to super-cars costing more than the average home. Due to the demands of the German Autobahn, even the most humble vehicle is capable of being brought from top speed to stationary in a respectable time and distance. Most modern cars, however, will object to doing this on a regular basis, which means that certain upgrades or modifications will need to be carried out for fast road, track or competition use. In this book I will guide you through some of the key operating principles of braking systems, the potential for upgrades, and choosing which will be most suited to your application.

SAFETY

Before starting any task, from a visual inspection right through to a complete strip and rebuild of the brakes, you must take stock of some important safety points.

THE CRITICAL NATURE OF BRAKING SYSTEMS

If you make a mistake during the install of something simple like a car stereo you may find it doesn't work, or not in the way you expect. Possibly the worst case scenario could be a fire. The best thing to do if you don't know what you are doing is to get in an expert, but you can mitigate most risks by simply preparing and doing research before you start.

If you make a mistake during the installation of braking components the consequences are suddenly much more severe. Although braking systems are on the whole quite simple, there are many tiny errors that could result in catastrophic failure. If you are not confident that you can carry out any of the tasks detailed in this book, then obtain competent assistance. There is no shame in asking for a fresh set of eyes to look over critical tasks, and in many industries, such as aviation, critical systems such as brakes usually require an independent check prior to being released to service after maintenance, just to ensure no mistakes have been made; after all, we are all human. If a seal is incorrectly installed on a brake caliper or master cylinder you could lose all pressure from the system, and on older single line set-ups this could mean a total loss of brakes. Don't be afraid to check your work.

TIDINESS IN THE WORK AREA

Although it sounds obvious, it is startling just how many otherwise very professional workshops are actually quite cluttered and untidy. This increases the risk of losing components, trip hazards and, of course, the ingress of dirt and grime into systems that should be kept clean. Keep a clean work area, especially when dealing with braking components. There should be two distinct phases to stripping components: dirty and clean. The dirty phase is when all the grime from the road is removed, old seals disposed of and contaminated fluid drained. Only when the component has been fully stripped and all contaminants removed can the clean phase start.

CHEMICALS AND BRAKE FLUIDS

In UK industry COSHH regulations (Control of Substances Hazardous to Health) have been in place for many years, controlling the use of various chemicals in the workplace. Most usefully for the home engineer, this has meant that comprehensive data sheets are available for the various fluids and chemicals that we use regularly, either online or directly from the product manufacturers. These allow the home user to check exactly what sensible precautions should be carried out when using harmful agents, and also give the best advice on what to do if an accident takes place. These data sheets, known as COSHH Assessments, detail the correct fire-fighting techniques, action to be taken if swallowed or the material touches bare skin, and other vital information. It is indeed very sensible to make sure you are fully informed about any substance you may be using, especially when it may be that you require some specific form of personal protective equipment (PPE) that might not be immediately obvious. Today it is very easy to protect yourself from harm with the amount of information available, and sensibly priced PPE is easily obtainable online. It is easy to minimize the risk to yourself: why risk harming yourself?

JACKING

Naturally, to reach the brakes of a vehicle, in most cases you will need to jack the vehicle up. The safest way of doing this is on a four post lift, but as that is beyond the means of most people we need to look at sensible alternatives. As we are working on the brakes it makes sense that either a single corner or an axle-set (i.e. both front or both rear wheels) will need to be elevated. First ensure you choose a safe area to do this: firm, level ground where there is no risk of the stand or jacks sinking or becoming unstable. Use high quality jacks and stands, and inspect them before use. Most items sold in Europe now come with a TüV marking, a statement that the design of the component you are using has been assessed by the Technischer Überwachungs-Verein (Technical Inspection Association). This is an independent group of organisations who assess equipment and processes to verify that they have been designed in a safety-conscious manner. If the jack or stand you have purchased has TüV accreditation then it has been assessed as suitable for its intended purpose *if used in conjunction with the manufacturer's instructions.*

All equipment, however, is capable of failing over time, through fatigue, misuse or damage. Inspect all lifting equipment before use, keep jacks topped up with the correct fluid, repair or dispose of any leaking equipment, and never use any item of lifting equipment that you have any doubts about. When jacking a single wheel on a vehicle, which has been suitably chocked to prevent it from moving, use either the standard jacking point on the car for that corner or choose a sturdy part of the vehicle sub-frame or chassis. It is easiest to slightly loosen the wheel nuts while the vehicle still has weight on the wheels. You should not jack on components that may deflect, become damaged or move in relation to the rest of the vehicle, such as steering arms or anti-roll bars. Ensure the jacking pad is not going to slip and is correctly designed for lifting a vehicle without damage. Remember that a vehicle raised too low will be difficult to work with, and a vehicle jacked too high will be less stable, so a suitable compromise must be found. Once the vehicle is slightly higher than the required working height you can lower it onto a suitable support. Do not use bricks as they can crumble, potentially causing the vehicle to fall. Ideally an axle stand with a very wide base should be used, capable of safely supporting the vehicle's weight while it is being worked upon. It is one thing to support a car's weight but entirely another to support that weight plus the additional forces to which it will be subjected while trying to undo stiff bolts or levering components.

BRAKE DUST

The most harmful materials that were formerly present in friction materials are usually absent from modern cars, although asbestos can sometimes be found even on relatively new vehicles. The dust generated by brake pads, however, is still not healthy to inhale. By its very nature the friction material is designed to degenerate slowly, in order that it doesn't break down rapidly during use, but the knock-on effect is that it will also hang around in your lungs for a long time. You can do some things to minimize the presence of dust, such as pressure-washing the area before you start, or damping down the area with a hand-held sprayer to keep the dust from becoming airborne. The best defence, however, is always to use the correct breathing protection to intercept the particles before they have a chance to enter your lungs.

BRAKE FLUIDS

Brake fluid is possibly the most flammable fluid in the entire car. As its very purpose is to operate in a high pressure pipeline it is often heavily compressed: a small leak will become a very fine spray capable of ignition purely due to its low flashpoint and the presence of hot materials in the engine bay. The exhaust manifold will easily generate enough heat to ignite a spray of brake fluid, with terrible consequences. Leaks in brake fluid systems are not dangerous just because of the reduced efficiency of the brakes, almost certainly leading to system failure, but because the fluid released has a high potential for causing a fire that could destroy the car. Ensure all unions and couplings are correctly made to a high standard, and any leaks are dealt with straightaway. Inspect systems for leaks regularly and never introduce any source of ignition when brake fluid is open or present.

Ensure you have a way of putting out brake fluid fires in the workshop: either CO_2 or dry powder extinguishers can be selected. Remember that blasting a fluid with a high pressure extinguisher could just be a way of blowing the burning debris around the workshop, so take care before squeezing the trigger. Ensure all brake fluids are stored safely; partly because of the risk of fire, but also because brake fluid is poisonous and absorbs water from the air once opened. Only buy as much as you need and dispose of any leftover brake fluid soon after it has been opened. While many fluids are unlikely to damage paint, it is still quite possible, so wipe away any spilled fluid immediately and irrigate with water. Always read the relevant COSHH data sheet so you know what to do if it comes into contact with your skin. This may vary, depending on exactly what is contained within the fluid.

TESTING AFTER WORKING ON THE BRAKES

The first run after the installation of a braking component or overhaul of the braking system is going to be crucial. This should be carried out on a dry day, so you can see any moisture indicating leaks. It will also allow you to test the braking ability on a road with the maximum friction, working the brakes as hard as it is safe to do so. Once the system has been fully inspected to ensure there are no leaks, the pedal should be depressed with the engine running to ensure the servo (if fitted) is giving maximum assistance. Press as hard as you would during an emergency stop; it is better that a weak seal or union fails now than once you are moving. After depressing the pedal hard a few times, stop the car and inspect every single disturbed joint, seal and union for any signs of leakage. Use a cloth to dab against the areas to more visibly show the presence of leaking fluid, and inspect the seals on the calipers for signs of ballooning.

Only then can a low speed road test be carried out where no members of the public could be harmed in the event of a brake failure; farmland or similar, with the landowner's permission, is ideal. By gently accelerating the vehicle a few times to a low speed, under 30mph, and then bringing it to a stop, you will be able to assess whether the vehicle is pulling up straight and true, and with a sensible amount of braking effort. If excessive effort is required, you may need to reassess your installation to confirm that it has been carried out correctly. Some pad materials do not work effectively while cold, or before they have been effectively bedded in. This means a certain period of on-road or on-track testing may be required before they can be said to be fully serviceable. Don't forget to check that the handbrake is working correctly as well; it is your backup in the event of a main braking system failure. Don't make this the last check you carry out on the system; after running the car for a few miles of use, repeat the same visual checks to ensure no leaks have developed, and check that the fluid level has not decreased. As the pads bed in there will be a slight drop in the fluid level, but this should be comparatively minor.

CHAPTER ONE

BRAKES: AN OVERVIEW

Braking, although based on firmly set scientific principles, seems to be shrouded in confusion and misleading information. Any highly regarded professional in the field, if asked whether there are any specific techniques one should use for setting up a braking system, tends to answer, 'suck it and see'. There are so many factors that affect how the brakes work: the effectiveness of your tyre compound, air temperature, ride height, damping rate, even just moving a component in the car can change the braking to a significant degree. Because of this complex interaction, getting an effective braking system to work at its best can, to a certain degree, sometimes be more of an art than a science. We can give ourselves the best possible chance of getting it right first time, however, by understanding the principles and how they make a difference to the way the brakes decelerate the car.

Brakes have an unusual potential for being 'sexy'; a number of times I have seen brake calipers fitted to vehicles purely for their aesthetic appeal, being completely unsuited to the application in hand. Proceed with caution: upgrades should be carried out only to the level that suits your needs. A vast increase in brake size will usually result in an increase in unsprung weight, with all the handling penalties that this incurs. Braking systems can also be costly, not only to install but to maintain. Consider carefully just how much braking you actually require. As a guideline, if you are not suffering brake fade and can easily lock the wheels, then you probably have enough braking for your needs.

Whether it is a friction block applied directly to the wheels or road, a six piston competition caliper on a ceramic disc or an air-brake, any system used to decelerate a vehicle is considered a braking system. The lineage of even the most modern systems can be traced quite clearly to the same systems fitted to the earliest automobiles, and this is because the basic principle was robust and simple in its application.

Braking can be considered an energy conversion process: the forward motion of the vehicle is turned, via friction, into heat. Heat is the primary limiting factor with most systems. The ability to dissipate heat is often what dictates just how much braking the vehicle can do. However, the factor of traction on the road surface is too important to ignore; the most powerful brakes are useless if they lock instantly without gripping to the road.

PADS

It is almost too obvious to suggest that pads should be the first port of call when upgrading a system, but when modern road cars make the transition to track cars, just changing the pads is sometimes the only modification required. Compounds designed for higher temperatures are available to suit almost every vehicle on the market and the compromises that high specification pads used to impose on the driver have almost entirely been eliminated. High temperature pad materials used to work only when the disc and pad was up to temperature, but most high spec friction materials now work very adequately from cold. Selecting a pad material depends on the weight of the car and the type of driving intended. Certain circuits, for example, are very heavy on the brakes, requiring frequent heavy braking and few opportunities to allow the brake assembly to cool. In this scenario a pad that operates at a high temperature is best, although it can be difficult to determine exactly what these temperatures can be. During testing it is possible to return to the pits and apply an infrared thermometer to the discs, but this only tells you what the temperature is once the car has already started to cool.

A more reliable technique is temperature sensitive paint, which is applied to the edge of the disc and changes colour depending on the temperature reached. A typical paint kit, for example, might give temperature ranges from 300 to 650°C. The paints would come in three or four different colours, depending on the temperature they represent, and would be applied in small strips around 1cm long. After a test session the paint is checked, and any colour that has bleached to white represents a heat level exceeded: for example, a disc painted with red, yellow and green paint returning to the pits with white, white and green paint remaining has exceeded the temperatures represented by red and yellow, which could be 400 and 500°C respectively. The unchanged green paint could represent 600°C, meaning you would need to find a pad material capable of resisting fade up to a maximum of 600°C. Some manufacturers actually incorporate temperature sensitive paint into their pad production process, coating the metal backplate in a paint that discolours when the peak operating temperature has been exceeded. Bear in mind that ambient temperature, track temperature, humidity and other variables may affect the readings; just because a pad material suits one circuit, it does not mean it will be suitable for others.

Pad wear rates vary drastically depending on the compound and use. For this reason, if you are using

Excessive wear due to the use of an incorrect pad compound, resulting in a melted piston and warped backing plates. ERIC JACOBSEN

a pad with which you are unfamiliar, it is prudent to monitor the wear more closely than usual.

Pads come in a confusing array of compounds and materials, with different levels of bite, wear resistance, heat resistance, dust production and noise level. A few of the materials used are listed here. The pad material is probably the most rapidly evolving part of the braking industry, so this list reflects only a very small section of a very complex area, and each manufacturer's pad is likely to consist of a complex blend of these products:

- Asbestos was used in brake pads owing to its excellent resistance to heat. It is rarely used today because of its carcinogenic properties, but can still be found on older vehicles.
- Aramid is a contraction of the term Aromatic Polyamide, a synthetic fibre used as an alternative to asbestos. The Nomex material used in race-suits is an Aramid fibre, as is Kevlar.
- Ceramic pads are perceived as quieter than their rivals since the frequency at which they squeal is usually beyond the frequency range of the human ear. They are generally more expensive than metallic pads, but kinder to the brake disc. They provide aggressive stopping power, good fade resistance and are lighter than metallic pads, making them ideal for competition. They tend to produce less dust, and the dust that is produced is lighter in colour and generally less noticeable. Because the temperatures generated are usually higher, and the pad does not conduct the heat away very effectively, a good quality disc should be used as it will be subjected to severe heat if used in competition.
- Metallic brake pads tend to be slightly noisier than the other types due to the metals used in their construction. They are more aggressive to the disc than organic pads, relatively heavy, and often used as standard equipment due to their lower cost.
- Organic is a term commonly used to designate a pad as being made of something other than metallic or ceramic compounds. Kevlar pads

are considered to be organic. They are usually softer and less hardwearing than their ceramic or metallic equivalents.

- Polyacrylonitrile is a synthetic polymer resin, and yet another modern alternative to asbestos.
- Sintered pads are a variation on the metallic pad where the metal compounds are fused together during the manufacturing process at a temperature below the melting point of the metals used. Other materials may be added to sintered pads that are not employed with metallic pads. These include compounds or materials added to alter their wear and abrasion characteristics. Sintered pads tend to sit above organic and metallic pads, but below ceramic pads in their braking power and fade resistance. They usually use copper as the base material bonding the others together, contributing to their increased cost over organic pads.

Some calipers use more than one pad, meaning different compounds of friction material can be used together to achieve a hybrid stopping ability tailored to the type of competition you are engaging in.

If you notice squealing or poor performance, especially from new pads, it is possible that the disc and pad may have glazed. Disc and pad glazing can occur when the resins binding the pad materials together collect on the surface of the friction material and crystallize, forming a smooth surface that does not grip the disc effectively. It is for this reason that most pads have a bedding-in period, which allows the excess resin to be boiled off. Manufacturers tend to have different bedding-in procedures depending on the intended use. A road car application, for example, might simply suggest that the brakes are only used gently for the first few hundred miles. A racing pad manufacturer might suggest a few high speed stops separated by long cooling periods, allowing the resins to be heated and then gaseously dissipate through the gap between the pad and disc before they get a chance to crystallize. No matter which procedure you use, the basic principle is designed to avoid constant application of heat and pressure, which will cause the resins to glaze. In the event that the pads do glaze up, it is sometimes possible to remedy the situation by gently removing the glaze from the surface with a mild abrasive paper. Do not apply any liquid solution to the pad other than brake cleaning solvent and lightly abrade without causing damage to the surface. In the event that the pad starts to break up or crumble, it should be discarded along with the rest of the pads in the set. The rotor should also be very gently abraded using a fine grit paper evenly across its surface, as the resin will have transferred to the metal. Do not use an aggressive machine for this; hand pressure with an even effort across the whole surface is the best technique. Rinse with brake cleaner to remove any contamination once you have finished.

Brake pads and other friction materials are assigned a code called the WVA number. The WVA number, an acronym of Waren Vertriebs Artikel (Merchandize Sales Numbering System), is generated by the Verband der Reibbelagindustrie (VRI, Federation of Friction Industries) and is used to provide a unique identification for the exact material and dimensions of the pad you are installing. Although it applies to all forms of friction material, the numbers relating to brakes are:

10000 to 14999	Car Drum Brake Linings
15000 to 19999	Commercial Vehicle Drum Brake Linings
20000 to 25999	Car Disc Brake Pads
26000 to 27999	Shoe Assemblies
28000 to 29999	Commercial Vehicle Disc Brake Pads

The intention is to remove the risk of human error resulting in the installation of an incorrect component to a critical braking system.

DISCS

The disc is one of the simplest components in the system and yet the manufacturing process is actually more complicated than you might expect. High

This BTCC car utilizes a floating disc. Note the temperature sensitive paint to denote when temperature thresholds have been breached.

SIMON CROSSE

quality discs are often dynamically balanced, and a great deal of design effort goes in to ensuring that the internal venting is efficient while maintaining a high level of structural integrity, something that requires a certain amount of compromise. Usually a disc brake is manufactured from cast iron (due to the ability of grey iron to conduct heat and withstand abrasive pad wear) and then machined to the correct tolerances. Often the only evidence of the original casting process is visible down between the vanes of the rotors. Any additional drilling or slotting will take place after the disc has been cast and machined, followed optionally by the dynamic balancing process.

The outside diameter of the disc will affect the amount of braking force applied for a given amount of pad pressure. The further from the centre of the hub that the pad is applied, the more leverage the pad and disc will be able to apply to the rotating mass, giving more stopping power, even if the total pad area has not increased. Therefore one upgrade that can be considered is to increase the disc diameter and move the caliper further away from the hub centre, using a custom bracket. The design of the bracket needs careful thought to ensure that it

will not allow flex or breakage under heavy braking. Ideally a high grade steel should be used, although aluminium can be used if weight is an issue, provided that the inherent weakness of aluminium is taken into account. The primary advantage of this modification is that the bulk of the braking system remains unchanged: there are no hydraulic modifications as the hydraulic system remains the same. Don't forget that improving the front brakes may throw the system out of balance, and that after moving the caliper the flexible pipe may be too short or foul something critical. The custom caliper bracket can be manufactured as a template out of MDF or a similar easily worked material: this allows you to test fit the wheel to ensure it fits over your new assembly and to verify that nothing is fouling. Under no circumstances, however, should you try driving the car using this template. The template can be used for both sides of the car as the bracket should be interchangeable.

For maximum efficiency the swept area of the pad should extend right to the very edge of the disc. Finding a disc to fit your car is often a matter of taking your original disc to a scrapyard or parts shop and physically comparing it with a few samples off the shelf. The new disc will need to have the same number of wheel studs, pitch circle diameter (PCD) and centre bore size, although this can be modified by a machine shop if required. The width of the disc should be as close as possible to the original to allow the pistons to retract sufficiently to allow the pads to be fitted, and should not be so thin that the piston extends beyond its design specification when both discs and pads are worn. Look at other models from your car's manufacturer as a good starting point for suitable discs to exchange. If the other models share components with your car it is even possible that other brake components will be interchangeable, meaning fewer custom parts need to be manufactured. VAG cars, such as Volkswagen and Audi, are a good example of this.

Make sure you consider the offset of the brake disc face from the hub face: this can be an advantage when making a custom bracket as it will give you more space, but also increases the risk of fouling suspension components or moving the caliper too close to the wheel spokes. Turn the steering from full lock in each direction to verify clearance at all steering angles, and ideally do this during different states of suspension compression. This is difficult to achieve on a vehicle with Macpherson strut suspension, but on a car with a double wishbone set-up the damper and spring can be disconnected to allow the suspension to be moved through all angles unhindered. While doing this, check that the flexible brake line does not become stretched or snagged by the new position of the caliper: the caliper may have been moved by only a few millimetres, but this can be enough to drastically change the geometry of the pipework. The pad should not overhang the edges of the disc in any way, with all of the friction material making contact with the bright surface of the disc.

Used discs can be used for the purposes of mock-up and testing, but new units should be installed for any road test and final use. If you need to modify the structure of the disc it is best to entrust this work to a professional machine shop. If you insist on modifying the disc yourself, however, always remove balanced amounts of metal (using a lathe, of course, rather than by hand) and do not remove any metal that could affect the structural integrity of the disc itself.

Sometimes a non-vented disc can be exchanged for a vented disc, but this will usually require a redesign of the caliper, a different caliper or a conversion kit. These kits will only work on a multiple piston caliper designed for separation into two halves, and usually consist of a spacer plate with the correct fluid galleries machined in, if required. It is possible to make these yourself, although a few safety measures must be considered due to the critical nature of their operation. Ensure you have manufactured the spacer accurately, with specific emphasis on the parallel alignment of the two mating faces. If in doubt, get them machined at a machine shop. They must be right. The width of the spacer should be precisely the difference between the non-vented disc and its vented replacement. If there are passageways for fluid in the caliper, ensure you use a gasket or sealant that is resistant to brake fluid of the grade you intend to use, and that there is enough surface area around the passageway to make leaks unlikely. Verify that it will take the pressure you will be applying, and also the heat: check this carefully against the manufacturer's specifications. If you are not getting fade using a non-vented caliper, it would be best to stick with it: if you are not going to benefit from the advantage of better heat dissipation, then there is no point putting yourself through the trouble of manufacturing modifications to fit a device that is heavier than the existing setup.

Slotting, drilling or otherwise venting the disc is another way of reducing fade, assuming you have

This Ferrari 458 employs carbon-ceramic discs with the calipers mounted close to the centre of the car, keeping the weight as close as possible to the ideal centre polar moment of inertia. If this car were front-engined, the rear discs would be smaller as more braking effort would be required over the front axle.

A two-part disc ready for assembly.
GOLDSMITH AND YOUNG LTD

The disc, showing the mounting holes by which it is attached to the bell. This disc displays little in the way of wear. GOLDSMITH AND YOUNG LTD

The bell, separated from the disc itself. As it is usually made of an alloy to reduce the unsprung weight, it is more easily damaged than a solid grey iron disc. The alloy also allows improved heat dispersion at the expense of complication and cost.
GOLDSMITH AND YOUNG LTD

already tried different pad materials. The other advantages of using a slotted or drilled disc include a very slight weight reduction, no additional modifications to the vehicle and, of course, the aesthetic appeal. The disadvantages include a reduction in the pad contact area, a decrease in pad life and an increased cost over a standard disc. The decrease in pad life is the result of the reduced surface area and the abrasive nature of cross drillings, which act like a cheese grater against the pad material. There is some debate as to whether this phenomenon is actually measurable, but so far the evidence suggests that it does detract from the overall life of the pad. Some manufacturers offer dimpled discs, where the disc is merely spotted with a drill rather than being completely drilled through, but this negates the advantages of the drilled disc in enabling the hot gases to exit the pad face via the cross drillings. If you are not having problems with overheating, then it is probably best to concentrate on other methods of dissipating the heat or using a pad material that is more able to cope.

Discs are available with exotic coatings to prevent corrosion. For most vehicles this will serve no purpose other than improving the look of the unswept surfaces behind the wheel. Within a very short distance any area swept by the pad will have its protective coating removed, leaving a region of metal that will rust, just like a disc without an

A Porsche wearing two-part discs with aggressive cross drilling. No fewer than six pistons operate within that large caliper. SIMON CROSSE

expensive coating. It is probably best considered as a gimmick for those who want to take their cars to car shows rather than to the track, as most track users will replace their discs fairly regularly anyway.

Corrosion on a disc will be taken off by the pad if it is fairly mild, for example if the vehicle has not been driven for a couple of days. If the corrosion is more severe it can aggressively damage the pads, reducing the contact area and having a severe affect on braking efficiency. The presence of salt on the roads can accelerate this corrosion and a new set of discs can be reduced to a terrible mess in a very short time. Rear discs are more prone to corrosion damage as they are less heavily applied, meaning surface rust sometimes doesn't get fully removed. It is possible to skim corrosion off discs, but only if it does not go too far into the metal and is unlikely to affect the integrity of the casting. Don't ever apply any product to the surface of the disc to prevent corrosion; anything that could lubricate the swept area of the disc is likely to be seriously detrimental to your ability to stop. The best way to prevent brake rust is to use the car frequently, and to store it in dry conditions. Manufacturer-applied coatings will last for only a short time at the beginning of the life

of the disc. Trying to 'drive-off' serious corrosion on discs will result in damaged pads. Either get the discs skimmed or invest in a new set. Remember when checking for damage to the disc to check both faces; the side you can see may be in a lot better condition that the side of the disc that is hidden.

Measuring the runout of a disc could give you a good indication of where a fault lies, such as a pulsating pedal. Runout is the amount by which the surface of the disc moves in relation to a fixed point of reference while rotating. A large amount of runout will give problems. It could be caused by anything from an improperly seated disc to a bent driveshaft. It is usually measured with a dial test indicator (DTI), a small gauge that moves its needle in direct proportion to the movement of a small plunger projecting from the side of the dial. The face of the dial can usually be rotated so that the zero marking can match the needle position as a reference before readings are taken. To measure the runout on a disc, the wheel is removed, the wheel nuts are replaced on their studs and then tightened to the correct torque to replicate the position of the disc during use. Spacer washers may need to be used if the threads bind before the disc is tightened down. The DTI is then mounted, using a clamped arm arrangement, to a stable part of the suspension that will not move in relation to the disc during the test. The tip of the plunger on the DTI is then brought into contact with the outer disc but within the swept area (around 10mm from the edge is usually sufficient). There should be enough preload to ensure it moves easily both in and out without coming to the end of its travel. The dial should be zeroed and the disc gently spun. The needle on the DTI will move slightly back and forth to show exactly how much runout is being generated. The maximum you would expect to see is typically 0.1mm total runout (i.e. maximum total deflection): any more than that and the disc is damaged or incorrectly seated, or a rotating component is bent. Acceptable runout values may vary based on the manufacturer's own tolerances, but 0.1mm is a good figure to work with if you don't have specific information for that particular vehicle. The test should be repeated on both surfaces of the disc, not just the one you can easily see.

It is sometimes possible to reduce excessive runout by matching the phase of the runout with the corresponding anti-phase runout of the hub, although this is only possible if the disc is not keyed to fit on the hub in only one way. The theory is that the hub will not display perfect runout, and neither will the disc, but the runout of each could be used at 180 degrees (or as close to it as possible) to cancel out each other. To do this, one of the wheel studs should be marked at its tip with correction fluid or similar; this is the reference stud. A corresponding mark is then made in the centre of the disc to line up the hub with the reference stud. Runout is measured using the above technique, and then the disc is removed and rotated by one quarter of a turn with reference to the hub for four stud cars, or one fifth of a turn for five stud cars (or a sixth of a turn for six stud cars, and so on). The process is repeated and the results recorded until the configuration with the least runout is discovered. This can often be enough to bring the system within the tolerances required.

Some manufacturers recommend the use of a brake lathe, which actually machines the disc on the hub, removing any runout issues as you make the hub suit the application. If this technique is used it is important to make sure the disc always goes back on the hub in the same place or the advantages may be lost. Brake lathes are often used on high performance cars, where the brake discs are expensive to replace, as a means of getting the most life from the disc. Rather than throw away an expensive component with minor scoring or corrosion, brake lathes allow the disc to be used again, assuming there is enough metal left in the disc to safely cut away the scored or damaged area. For vehicles where the disc is much cheaper the cost of the machining process is not normally a viable option, but for performance cars, with discs costing many hundreds of pounds each, it can be a useful saving. Make sure you use a reputable operator who will not machine the disc beyond its safe tolerance, as stated by the disc manufacturer.

CALIPERS

When talking about a brake upgrade, most people immediately latch on to the brake caliper as the most important thing to change. As mentioned above this is not necessarily the first port of call. If you are still not getting the required results after upgrading the discs and pads, it may then be time to look at alternative calipers. Aftermarket kits are available for most cars, usually matching the aftermarket company's generic caliper with a bracket that makes it suit the specific application. Often referred to as 'Big Brake Kits', most use the opportunity of the caliper upgrade to move to a larger disc as well. If this is your intention, make sure the intended upgrade will fit under your current wheels or budget for a larger set. If you intend to upgrade to a 'scrapyard' caliper, then the same process of bracket manufacture should be followed as was discussed above. Of prime importance is the matter of friction material overhang, which should be avoided wherever possible. Some people even advocate trimming the friction material from the backplate, but this leads to all sorts of problems,

A three-piston caliper stripped prior to rebuild. Note the cleanliness of the internals; this is vital when dealing with the hydraulics of braking.
GOLDSMITH AND YOUNG LTD

including the risk of de-lamination of the pad. If the pad doesn't fit, look at the other pads designed to fit in that caliper, as calipers are often fitted to a variety of disc sizes across manufacturers' model ranges. If there are none fit for your application, then either the disc or the caliper is not right for the application and must be changed. Your goal is the maximum swept area as close to the edge of the disc without going outside the disc diameter; any other configuration is substandard.

A few things need to be considered when selecting an aftermarket caliper. Cost is one of the first to spring to mind: what sort of advantages will be gained from fitting this caliper against the price of an upgrade? Spares availability is another. Supporting a small business is laudable, but will the pads for their caliper still be available should they go out of business? And does a low volume caliper have the varied support from friction materials companies that may be had from a more mainstream company? Nissan calipers are a common upgrade in modified car

circles as their 4 pot calipers can be purchased second-hand relatively cheaply. Look closely, however, and you will see there are two variants: aluminium alloy and cast iron. It goes without saying that the cast iron option is much cheaper to buy second-hand as the additional weight makes it less desirable: the caliper material can have a large bearing on just how useful the caliper will be to you. If you are upgrading the rear calipers you must also think about the handbrake mechanism. Sometimes it is more convenient to install a separate mechanical caliper specifically for this, although this introduces extra weight and bracketry. More usually a caliper specifically designed for use on the rear of the car is the most elegant solution. Changing the caliper can affect the anti-lock braking system (ABS), as the amount of pressure and fluid it is required to supply and release during operation may vary from its design specification. Trial and error is often the best way to find out if this is the case: the system will most likely show a fault code should a problem be present.

CHAPTER TWO

DISC BRAKES AND CALIPERS

BRAKE CALIPERS: A BRIEF HISTORY

Due to the simplicity of manufacture, the brake drum was the primary means of retarding a moving vehicle even as late as the early 1980s. A few American manufacturers toyed with the disc brake system, notably Chrysler, but the first patent was actually held by Lanchester as early as 1902. The Lanchester used copper brake pads, making them incredibly noisy and unpleasant to use. Due to the poor quality of friction materials available, they never really caught on until Dunlop first employed them on the Jaguar C-Type racing cars. This technology soon found its way on to Jaguar's road cars (although Triumph's TR3 was the first production car to sport discs, and the Austin Healey 100s was the first to feature discs all round). The motoring press initially referred to them as 'plate brakes' before the term 'disc brake' caught on. The improvement in braking was deemed so marketable that Jaguar thought it prudent to warn other road users, boasting of better stopping power with badges on the rear of their saloons so that other drivers would give them more space (and advertising their cars' superiority over their competitors probably wasn't far from their minds . . .).

In reality, however, a well-maintained drum brake was a match for early discs; the superior fade resistance wasn't really of much concern to traffic in everyday motoring. What was useful was the reduction in required maintenance; drum brakes required regular adjustment to keep them operative effectively and could be difficult to keep balanced. Neither was an issue with disc brake systems. Although the pads on disc brakes had much less surface area, which in theory led to more rapid wear, in practice they proved to be much easier to live with. Racing teams found the reduction in unsprung weight to their liking, as well as the increased fade resistance. Drum brakes tend to retain heat due to the large surface area of both the friction material in contact with the metal, and the size and design of the cast iron drums. Some efforts were made with racing drum brakes to add fins intended to aid both heat dissipation and add reinforcement, but the cup-shaped design of a drum brake meant airflow was difficult to channel to the important operating components. The open-faced disc brake was more suitable for ducting air to, and its simple design meant that the heat generated was soon shed into the air (as well as allowing the highly abrasive brake dust to fall away rather than building up to cause harm). This open design made the disc brake better able to shed water during rainfall or passing through flooding; the drum brake was slow to do so and could display much compromised braking characteristics until the water had dispersed. In 1967 Porsche became the first manufacturer to assist the cooling of their 911's disc brakes by using an air gap between the two metal faces. These ventilated discs showed further improvement by physically displacing the air from the centre of the hub with the internal vanes in the casting. These acted like a centrifugal fan, causing air to be drawn in through the centre of the disc and ejecting it along the circumference; a simple modification with measurable improvements.

Disc brakes were primarily used on the front wheels due to the increased work the front braking axle is required to do. As they also tend to require greater mechanical effort to operate than a drum brake, the drum is still often retained on the rear of vehicles to make handbrake operation simpler and easier to operate. As drum brakes require the hydraulic components to move further to operate they are also less prone to seizing in lightly loaded applications, such as on the rear wheels of small domestic vehicles. Where discs are used on the rear axle the handbrake mechanism can be considerably more complicated, becoming one of the more troublesome parts of the system. Some manufacturers used inboard disc brakes, notably on earlier Jaguar cars. This provided excellent unsprung weight characteristics but moved the brake to a place that is hard to access, next to a hot and often oily differential. Inboard discs can provide additional complication in that the driveshaft has to be removed to change the disc.

Drum brake designs are easier to operate, as with careful design they can be made to assist the operator and reduce pedal effort. A form of mechanical assistance was provided by aligning the brake shoes so that the rotation of the drum would pull the brake shoe on more firmly. By comparison disc brakes require a lot more pedal pressure, which necessitated the fitment of a servo on most disc-braked cars. The servo increases the amount of mechanical effort applied to the hydraulic system for a given amount of effort applied by the driver. Although now standard on all disc-braked cars, they were a popular accessory in the 1960s and '70s, and could be retrofitted by the keen motorist to existing vehicles using either drum or disc brakes.

Brake calipers come in single or multiple piston varieties. Most units until the mid-1980s were of a twin piston design: the pistons faced each other and squeezed the disc from each side. As they are hydraulically fed from a common brake line, the amount of pressure applied tended to be very even. Modern mass-produced vehicle braking systems tend to use single piston sliding caliper arrangements. This means that, rather than rigidly mounting the caliper, it is allowed to 'float' on greased sliders. A single piston operates against a slider, between which the caliper is clamped. Due to the floating of the caliper the applied force is even to both sides of the disc. The advantage is that hydraulically the system is simpler, thus being cheaper to produce and repair. The disadvantage is that these units require the sliders to be in good working order to provide even pad wear, and they often fall into poor repair due to the inherently low-maintenance nature of discs. High performance calipers might use four, eight or sometimes more pistons. Twelve piston calipers, six pistons on each side, have been introduced. Using more pistons with an individually smaller radius means the pad material

can be larger and follow the shape of the disc further, but the additional expense and complication means these systems are expensive and are largely overkill for most applications.

Alternative disc materials came from unusual sources. Carbon discs and pads were first used on aircraft and have now found their way into racing. High performance cars and race cars increasingly use ceramic discs, a technology that came, rather surprisingly, from the rail industry. The high speed trains that were being dreamed of in the late 1980s made very heavy demands on braking systems, requiring extremely heavy rolling stock with relatively few wheels – and therefore relatively few points to which braking could be applied – to be brought to a halt safely in a short distance from speeds in excess of 320kmh. The excellent heat resistance and dissipation that ceramic discs displayed has shown potential in motorsport.

When carbon discs are being discussed the material in question is actually carbon fibre reinforced silicon carbide. This makes for a light but high friction material with an excellent resistance to wear. The disadvantages for road cars are the very high purchase cost combined with poor 'feel' at lower road speeds. This is offset by their excellent high speed stopping capability, relatively low wear rate, corrosion resistance and dust that is less aggressively corrosive to wheels and paintwork than the standard braking system equivalent. The high cost is primarily due to the length of time it takes to construct the disc; whereas a cast iron disc can take a few hours to make from start to finish (taking into account casting, cooling, machining and so on), a carbon ceramic disc can take a week or more due to the labour-intensive manufacture and inspection process and the lower production volumes. The hefty price may be considered quite reasonable in high-end motor sport when the light weight and its heat resistance, being capable of tolerating temperatures in excess of 2,000°C, is taken into account. One unusual characteristic of carbon brakes is the way in which they wear. One heavy application of the brakes will typically cause less wear to the rotor and pads than a few lighter applications, as it is the initial bite that causes degradation to the friction material. Carbon discs require a specific pad material, usually a complementary but softer carbon composite, which is designed to wear at a quicker rate than the rotor.

REBUILDING A CALIPER

Brake calipers are a highly stressed part of the braking system, undergoing extremes of heat while maintaining a hydraulic seal on a system that contains a relatively high pressure fluid. A further complication is that brake fluid is easily the most flammable liquid carried in a car, and yet it is expected to operate in the proximity of components that can glow red hot in use, all while being routinely soaked in freezing water and caked with abrasive dust from the brake pads. When you consider these assaults it is amazing they are as reliable as they are. Because of this, when they do go wrong it is tempting to just buy a new one. They are not complicated to rebuild yourself, however, and a self-rebuilt caliper gives the added satisfaction of knowing the work has been carried out to a high standard. The following description assumes we are dealing with a single piston caliper: if the caliper contains more than one piston, simply repeat the relevant sections on each piston. If you are rebuilding one caliper it is not much more effort to rebuild the other side's caliper as well; this improves your chance of getting well-balanced braking. Calipers can be made of cast iron, steel or lightweight aluminium alloys. You should be aware that the alloys are softer than their cast iron equivalent and will not take much physical abuse, but as well as their weight advantage they are also capable of dissipating heat more readily, making them more suited to racing applications.

PREPARATION

There are two halves to the job, and they need to be kept separate. Disassembly and assessment is a grubby task, whereas reassembly needs to be as clean as possible. Set yourself an area of workbench or table that can be cleaned to a high standard, as any contamination that gets into the hydraulic passages could be very detrimental to your braking.

Required Parts

- New rubber seals, sliders and, if required, piston
- Sockets and spanners to remove the caliper from the car
- Brake line clamp (a set of mole-grips and a couple of large sockets can suffice)
- Brake union spanner
- 2000 grade wet/dry paper
- Dental pick (recommended but not essential)
- Latex gloves, dust mask and goggles
- Airline or foot-pump (recommended)
- Brake fluid of the same grade you intend to use in the car
- Brake cleaning fluid
- Brake rubber grease
- G-clamp
- Cleaning tray
- Stiff brush
- Toothbrush
- Paper towels

SAFETY

Brake fluid is normally pretty nasty stuff. Even worse is used brake fluid, contaminated with gritty brake dust. This makes a nasty combination you don't want to get on your skin or in your eyes. Even when it doesn't contain asbestos, which was still found in new cars as recently as the early 1990s, brake dust is pretty unpleasant and should not be inhaled. Use latex gloves or similar to minimize skin contact, and a dust mask for any job where brake dust may get disturbed. Eye protection is also advised, especially where brake fluid is likely to be splashed. Brake cleaner is, by its very nature, an aggressive solvent. Gloves and goggles should be used, and any manufacturer's safety precautions must be followed. Remember that brake fluid doesn't do your paint any favours; if any gets on your bodywork, wipe it away without delay.

- Brake assembly grease (specifically for brakes, it doesn't attack the rubbers or react with brake fluid; it is not the same as copper grease)
- Copper grease
- Brake fluid of the correct grade
- New pads
- Pipe and fluid receptacle for bleeding (must be resistant to brake fluid; a clean jam jar is fine)

REMOVAL FROM THE CAR

Each car is different, but broadly speaking, removing the caliper from the car involves safely jacking the vehicle and placing a secure axle stand under the corner you will be working on. Once the wheel is off you will have access to the braking system. Normally the disc can be left in place. Place the brake line clamp on the flexible rubber hose between the body and the caliper. If you don't have a brake line clamp, which is inexpensive and well worth investing in, you can slip some large sockets over the jaws of a pair of mole grips. Remember when applying any clamp to the pipes that you are only trying to close up the space in the middle; you don't need to squash the pipes too hard. Any heavy deformation of the

The caliper before cleaning.

rubber pipe will damage it, and any damage to these pipes renders them scrap. Don't take chances; these are critical to your safety. Use this as an opportunity to inspect and replace the hose if any damage is present. No damage is acceptable in this case: renew if in any doubt.

With the pipe clamped you can now disconnect it from the back of the caliper. If possible use a brake pipe spanner, since brake unions are hollow and collapse easily. A brake pipe spanner gives you full contact around as much of the union as possible, reducing the risk of damage or rounding. Cover the end of the removed pipe to prevent dirt and moisture ingress: the finger of a disposable latex glove is often ideal, secured in place with an elastic band or small cable tie. Do not remove the pipe-clamp until the caliper is reinstalled. Use this opportunity to loosen the brake bleed nipple, as it is easier to loosen while on the car. Unbolt the caliper, usually retained with two large bolts accessible from the rear of the unit. It is advisable to take the caliper from the car with a piece of paper towel held over the hydraulic ports to prevent fluid spillage.

STEP ONE

Place the caliper and its carrier in the tray and scrub using the stiff brush and brake cleaning fluid. If you have to remove anything, don't forget to make

---

INSIDER TIP

People often talk about how braided brake lines offer significant improvements over standard lines. However these are not without their issues: you cannot use hose clamps with them, so interruptions to the system require a re-bleed every time, internal damage is not immediately obvious, and unless they are installed with care they can rub against bodywork, suspension components or even the tyre. The primary advantage is that the external reinforcement of the pipe reduces the amount of expansion, which means more of the brake pedal force is transmitted to the pads. The difference is often too small to notice, so it is up to you to decide whether the advantages outweigh the disadvantages. Personal preference will normally dictate this choice.

Knocking out the pad retaining pins. These would usually be replaced as a matter of course, but they can be inspected for straightness and wear and reused, if safe to do so.

Removing the pads. This is much easier when the caliper is still fitted to the car.

careful notes of how it was fitted, and lay it out in an area on the bench where it can't easily be dislodged. Don't remove any seals at this stage, but try to get off as much grime as possible.

INSIDER TIP

If you think you might be in any doubt as to how the unit will go back together, grab your camera. A few photos now will make life a lot easier when you are later trying to reassemble a complicated unit.

STEP TWO

Dry off the caliper with the paper towels. Take this opportunity to assess the state of the outside of the unit. If any cracks are visible or the hydraulic threads are stripped, the unit is scrap and cannot be used again. Put the caliper in your vice, holding it on the mounting lugs only. Never clamp the caliper around the cylinder or you risk distorting it, especially with softer alloys. If there are any handbrake or adjuster mechanisms connected directly to the piston, these should be examined and removed before attempting to force the piston.

Using the airline or foot-pump, pressurize the cylinder chamber until the piston pops out. An airline with a rubber-tipped gun will allow you to make a good seal against the port. An old but serviceable master cylinder mounted to a handy frame, fitted with a lever so that it can be operated easily, makes life a lot easier, especially if stripping

Undoing the pipe to the caliper. A brake pipe spanner drastically reduces the risk of damage to the union.

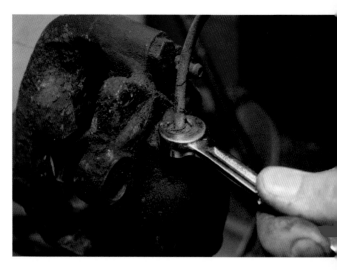

Draining the caliper over a suitable container. Remember that brake fluid is harmful to health and the environment.

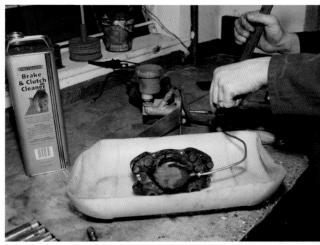

The pistons being pumped from the caliper using a home-made rig. This is worth manufacturing as it is easily the safest way to remove pistons.

The union removed earlier. Make a note of the shape to ensure the correct replacement is sourced.

unit has a link pipe between two or more cylinders then carefully remove it. Assuming it is not damaged it can be used again. If it is corroded or cracked, or the union threads are suspect, then a replacement can usually be made using copper or Cunifer pipe.

INSIDER TIP

You might need to use some ingenuity to improvize a connection between the pump and cylinder. If you struggle you can, as a last resort, reconnect the caliper to the car and pump the piston out with the pedal. Be aware that the system will need bleeding fully afterwards, which on some cars can be a full day's work. Never push the pedal all the way to the floor as you risk damaging the master cylinder seals.

---

calipers is something you do regularly. It's a good idea to put something in the caliper to protect both the caliper and the piston in case the piston leaves its home with any speed: a piece of scrap wood is ideal. Keep your fingers well clear. Take this opportunity to remove the old sliders. Don't throw them away just yet, but compare them alongside the new parts to ensure you have the correct items. If the

STEP THREE

Assess the cylinder and the piston. Minor scuffing can be polished out using wet and dry paper, but anything deep will render the caliper unusable. If there is scratching to the piston there will almost certainly be damage to the cylinder wall, in which case just replacing the piston is not enough to make the unit serviceable. Using the dental pick or a similar hooked removal tool, very gently remove the rubber seal from the cylinder wall. If there is more than one seal, make sure you remember where each one came from. Try not to damage the seal: even though you will almost certainly be throwing it away, it may prove useful if you need to size up a replacement.

Even more crucially, do not scratch the inner wall of the cylinder or you will increase the likelihood of a leak on reassembly. Once all rubber components have been removed the final clean can be carried out, using brake cleaner to remove all traces of any dirt, residual rubber or brake fluid. You can only be satisfied that this step is complete once the caliper is perfectly clean.

The pistons forced out. Make a note of the arrangement of the dust seals.

INSIDER TIP

A cable tie end makes a useful tool if you are worried about scratching the wall of the caliper while removing the seal. It is more difficult to use than a metal tool like a dental pick, but the chances of damaging the wall of the caliper are negligible.

STEP FOUR

Clean up: the work area needs to be completely spotless. Good practice is to line the work area with clean paper: newspaper is perfectly adequate for this. Take this opportunity to clean your hands and put on fresh gloves.

Removing the bleed nipple using a hexagonal socket to reduce the risk of damage to the flats.

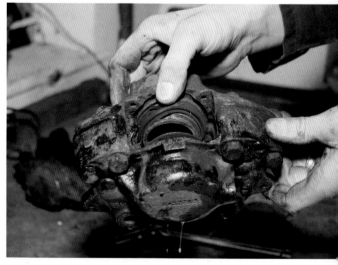

These rubbers are damaged beyond repair, but they should be replaced regardless of condition.

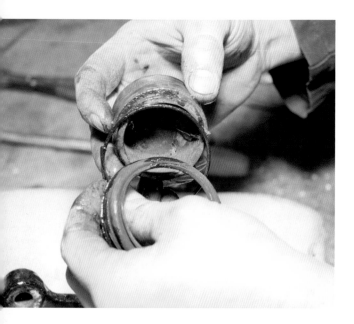

The condition of the pistons can be assessed once they have been removed. It is usually economically viable to replace them regardless. These are past their prime.

STEP FIVE

Use a little wet and dry to polish out any scuffing. It is important not to concentrate too heavily on one area of the piston or cylinder because this will give an uneven shape and increase the risk of seal failure. For a much more professional result a hone can be used to polish out any imperfections. Be aware that it is possible to accidentally bore the cylinder over-size, so only carry out the minimum honing required to remove scuffs or marks. Any lips or sharp edges that should not be there will damage the rubbers in use, reducing their life accordingly. Corrosion that is sufficiently deep to have pitted the piston is usually serious enough to require replacement. Once any scuffing has been cleaned off the piston, the work area should again be cleaned to a high standard, leaving no residue of the abrasive compounds present in the wet and dry. You can use an airline to blow out any ports that might have traces of abrasive or old brake fluid, but wear goggles and be careful not to spray the abrasive/brake fluid mixture over anything or anyone.

STEP SIX

Once you are content that everything is clean and the seals have been removed, try fitting the cylinder back into its bore as a test. It should slide easily without the seals fitted; if it doesn't then further investigation is required. Look for any high spots caused by corrosion or distortion of the piston itself. If the piston is distorted (unlikely but not unknown) then a replacement will be required. Any high spots that are minor can be polished back with wet and dry. Anything major will again require a replacement piston.

Using a high torque nut gun to undo the caliper bolts holding the two halves together. From this moment cleanliness becomes much more important.

The mating faces in this example are quite clean. Damage here, although rare, is often enough to scrap a caliper.

The seal between the fluid galleries in the two halves of the caliper. Treat this with extreme care, as the component is seldom changed and it could be difficult to source replacements on some older vehicles.

Removing the old outer seal.

Removing the old inner seal.

Gently removing the seal to minimize damage.

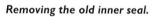

This badly corroded piston could cause damage to the bore on removal. Use clean brake fluid as a lubricant to ease it free.

A dental pick or similar is a useful tool for removing the more stubborn dirt. Take care not to damage the surface of the caliper, especially where the seals are concerned.

Rinsing the components of any traces of brake fluid or grease prior to grit-blasting.

Mild corrosion can be lightly polished out using a fine grade emery paper.

An airline is useful for blowing out any traces of dirt from the internal galleries. Always use correct eye protection.

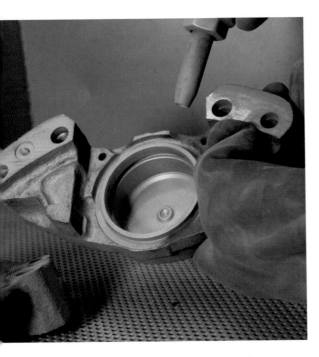

If you decide to grit-blast the calipers, make sure every trace of the grit-blasting media has been removed before painting, plating or reassembly.

Mating faces were treated with care to avoid issues with sealing later.

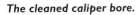

The cleaned caliper bore.

As before, a quick blast with the airline will dislodge anything left inside the caliper.

Very little effort . . .

. . . yields some pleasing results.

Polishing the caliper before plating.

INSIDER TIP

If you are keen to paint your calipers, then now is a good time. Use a paint with a high temperature resistance, and mask off anywhere that requires components to slide over each other. Blank off any hydraulic ports and tape up the cylinder. Painting a caliper usually serves no purpose other than cosmetic, and can sometimes cause issues by making the caliper stick if paint is allowed on surfaces where it shouldn't go. Remember also that if you need to change one caliper in the future you will have a mismatched set.

STEP SEVEN

If you are content that the caliper is spotless then reassembly can begin, although it is also a good opportunity to grit-blast them and have them sent off for plating. Assuming that this has been done, and all traces of grit-blasting media removed, you can proceed with assembly. Ensure that no adjusters or handbrake components need to be fitted during this process. At no point should excessive force be used. Lightly lubricate the inside surface of the cylinder and the seals with brake assembly grease. Ease the seal into place, being careful not to damage the rubber in any way; repeat this as required if more than one seal is present. Insert the lubricated piston, being very careful not to nip the rubber seals, but don't push it in all the way just yet. Using a little more of the grease around the inner edge of the dust seal, stretch the dust seal over the end of the piston. Take care not to damage it. You can now ease the piston in further, easing the outer edge of the dust seal into place when the piston is sat in the correct position. Once this is done, gently ease the piston fully home; this is easier done now than on the car once fluid has been introduced. Now is as good a time as any to refit the bleed nipple.

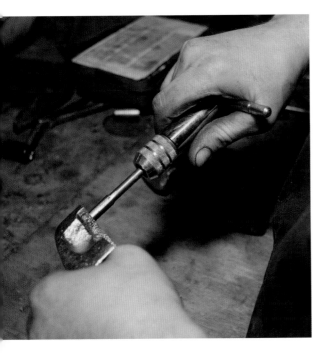

Cleaning the threads with the correct size tap.

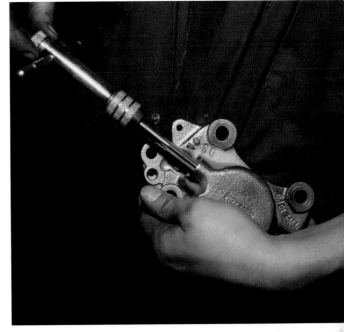

Repeat the tap process with the bleed nipple. Make sure you are not cutting away metal, just removing any build-up in the threads.

The plating and cleaning process has left plenty of detritus in the threads.

Using a drill bit with the tip cut to the correct angle for the application, gently face the seat by hand.

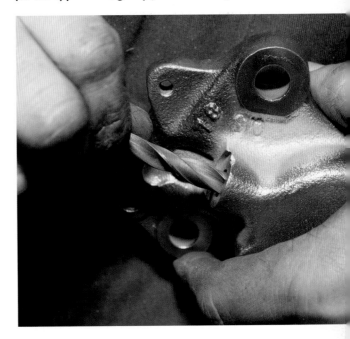

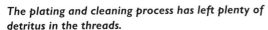

The dental pick comes out again, this time cleaning the bore before installing the rubbers. Cleanliness cannot be overstated.

A clean work area ready for reassembly.

STEP EIGHT

The caliper can now be reassembled, using new sliders where required. Take care to keep any dirt out of the hydraulic ports; placing a piece of tape over them can help, but it should be large enough that it doesn't get forced into the port accidentally during installation. You are now finished with the brake assembly grease: if anything requires further lubrication, use the copper grease sparingly instead. You don't want it to cause clumps of dusty grease and you certainly don't want it on any friction component, such as the disc or pad. A thin smear on any surfaces where metal will move over metal, such as the pad guides, is usually sufficient.

Lightly honing the bore to improve the seal, removing as little as possible.

Drying out the ports, and finally we are on to assembly.

All components laid out for inspection prior to rebuild.

A final rinse with brake cleaner.

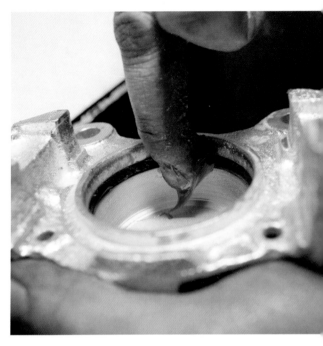

Lubricating the seal before installation. Note the use of braking system-specific assembly lubricant.

A thin smear of assembly grease will ease piston fitting and reduce the risk of damage.

First seal in place.

A brand new piston also gets a thin layer.

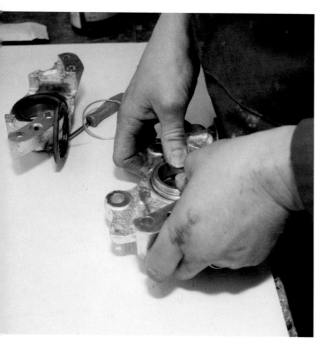

Easing the piston home. Make sure you keep the piston parallel with the bore.

Take care not to pinch the rubber during fitting.

Now the outer seal or dust cover can be installed, using a metal retaining clip.

A well-installed piston with rubber seals and metal retaining clips in place. Repeat for each piston.

The two halves can at last be reassembled.

The two halves mated. Check the mating seal is correctly seated before tightening the installation bolts. Some specialists recommend replacing these bolts every time the caliper is separated.

The rubber mating seal is lightly lubricated before fitting.

Bolting the two halves together using a hexagonal socket to minimize damage.

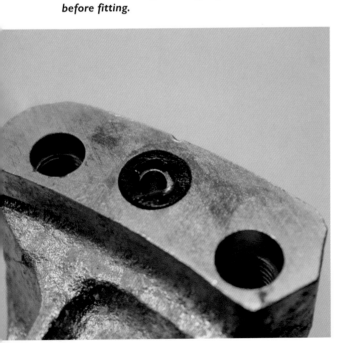

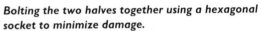
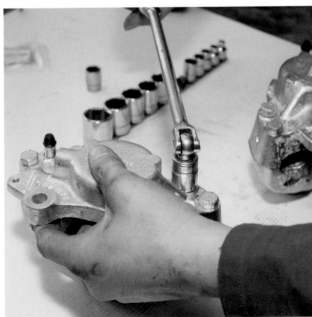

Ready for installation. Remember to bleed and check for leaks before driving.

Here a torque wrench should be used to the manufacturer's recommended figures. When these are not available you should choose the tightness based on the type of metal used. Alloy calipers will strip their threads more easily than a cast iron caliper.

One last clean before fitting to prevent dust sticking to the brake assembly grease.

STEP NINE

The caliper can now be reinstalled on the car. New pads should be used on both sides, even if the caliper on the other side was not rebuilt. The disc should also be changed if it is coming to the end of its life. Unless specifically stated otherwise by the manufacturers, a thin film of copper grease on the back of the pad will help prevent brake squeal. Once mechanically installed and the hydraulic system reconnected, the brake pipe clamp can be removed.

STEP TEN

The hydraulic system will now need to be bled. Any residual air in the system will cause sponginess in the brake pedal, so it is important to evacuate it completely. Modern systems with ABS sometimes need additional electronic units to put the ABS pump into a maintenance mode that displaces any air trapped there, but this should not be necessary as we have interrupted only a small part of the system. Ensure that the brake fluid reservoir is full of the correct brake fluid and then lightly replace the cap without tightening it down; if you leave the cap off entirely, the brake fluid can splash when the pedal is depressed. Keep an eye on this reservoir throughout the bleeding process as running it dry means you will have to re-bleed the whole system.

At this point you will need to draft in a helper to press the brake pedal while you operate the bleed screw. Make sure the helper can hear you clearly and that you have established a clear technique for communicating what you want them to do. One suggestion is to say 'down' when you want them to pump the pedal down, and for them to respond 'down' once it has reached its low point. When you wish them to release the pedal the same process is applied using the word 'up', with the helper repeating this once the pedal has been returned to its rest position.

With your helper in place, attach a ring spanner of the correct size over the bleed nipple and put the pipe into the jar. Opening and closing the bleed nipple should be done with care, never over-tightening or using undue force. Remember that the nipple is hollow and will shear easily. Loosen the nipple and then nip back up enough to just close it. Open the nipple as your helper presses the pedal down slowly and smoothly, nipping it shut again once the pedal is near the floor. Advise your helper not to go fully to the floor as the rubbers in the master cylinder can be damaged by excessive travel. Once the nipple has been shut the helper can smoothly return the pedal to the rest position. Watch the fluid coming from the nipple: in the early stages of bleeding it may take a few pumps to get anything from the system. Recheck the fluid level in the reservoir every few pumps and top up as required. The nipple can be tightened for the final time only once clean fluid is coming through without any air pockets or bubbles; it should be tight enough to seal, but not so loose that it can come undone. Remove the drain pipe and wipe up any spilt fluid.

INSIDER TIP

Racing teams often use two different types of fluid that are identical in every way except for their colour. By alternating two heavily contrasting colours, such as blue and yellow, you can easily tell when the old fluid has been purged from the system. Using a bleed pipe with a non-return valve can also make it easier to bleed the system, but be aware these valves are not always as good as they first appear and can often fail after a few uses. Check them before you put your trust in them.

STEP ELEVEN

Your helper should now apply as firm a force to the brake pedal as would be applied during an emergency stop. Wearing eye protection in case of high pressure leakage, inspect around the caliper, looking specifically for any leakage of fluid. If no leaks are visible then the wheel can be refitted and the car lowered to the ground. A test drive should be undertaken to see that the car pulls up straight under braking.

CHAPTER THREE

DRUM BRAKES

The drum brake is in many ways more complex than a disc brake, as it requires a mechanism for adjustment and a manner of retracting the brake shoes, neither of which are considerations for a simple disc-based system. In defence of the drum brake, the hydraulics are less prone to seize as the pistons move further during use, they require less pedal effort and therefore can be operated without a servo, a handbrake can be incorporated very simply and the manufacturing processes require less stringent tolerances. Many older vehicles and those manufactured with a budget in mind still use drum brakes, so they remain important and worthy of discussion.

The drum brake consists of a cast metal (usually iron) drum that rotates with the wheel. Inside this drum, mounted to the axle assembly, is a pair of shoes that follow the internal contours of the drum. In normal use the shoes are held away from the drum with springs, but on applying brake pressure

the shoes are forced outwards by hydraulic cylinders, which overcome the return springs and make the shoes contact the inside of the drum. The shoes are normally operated at one end by the hydraulic cylinder and pivoted at the other. Some brakes place both pivots at the same end of the drum, meaning only one hydraulic cylinder is required. This system is self-adjusting to a limited degree as the cylinder usually floats and centralizes itself during operation. This arrangement is equally effective when the vehicle is moving forwards or backwards. On a car with drums on each wheel, or discs on the front, it would generally be used on the back axle. Another arrangement is to give each shoe its own pivot point and individual cylinder. This means the shoes can be arranged to use the rotation of the wheel effectively to assist with their application; the direction of rotation is the same as the axis of shoe application, so the drum pulls the pad material harder into the drum's inner wall. In reverse, however, this means

A leading shoe/trailing shoe drum brake.

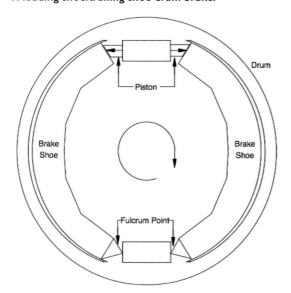

Twin leading shoes.

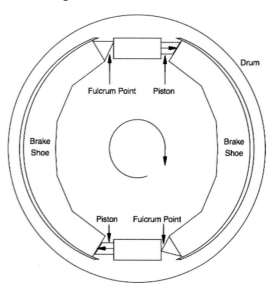

the shoes are less effective, as the opposite rotation acts to reduce the braking effect. As a result these brakes are more likely to be used on the front axle, where the servo effect would be more advantageous, and a standard single-cylinder arrangement on the rear provides effective braking in both directions of travel. The arrangement of the brake shoe with relation to the direction of travel gives it the designation leading or trailing shoe: a leading shoe operates in the same direction as the wheel rotates; a trailing shoe operates in the opposite direction to rotation. Therefore drum brakes can be described as twin leading shoe (for a two-cylinder system used on the front wheels) or leading and trailing shoe (for a single-cylinder system most likely used on the rear).

ADJUSTMENT

Older drum brakes are usually adjusted by means of a mechanical cam that takes some of the free movement out of the system, whereas modern systems use an automatic adjuster. Additional free movement will appear in the system as the shoes wear; this is usually noticeable as increased pedal travel. With a leading/trailing shoe arrangement the adjustment is usually a simple cam placed between the two pivot points of the shoes. By rotating the cam the pivot points are forced further apart, meaning less travel is required at the opposite end of the shoe where the cylinder applies the braking effort. This system applies an equal amount of adjustment to each shoe: providing the wear has been equal, it is very easy for a novice to carry out. The adjuster can usually be accessed either from the rear of the drum or through a hole in the face. The technique for adjustment is usually to spin the wheel and check for freedom of movement, then adjust the brakes until the adjuster goes tight (the adjuster should usually be turned clockwise to tighten, although this is not always the case). Spinning the wheel should now result in drag from the brakes being applied with no pedal pressure. The adjuster should then be backed off until the wheel spins freely again with no drag. It is best to apply the pedal before spinning the wheel to re-centre the shoes.

An alloy finned drum on a competition spec Mini. The fins assist with cooling and give additional strength. ADAM GALLAWAY

Once the brakes on both wheels on the axle have been adjusted, they should be checked to ensure they are capable of stopping the car from a low speed before braving the highway to see if the adjustment has been successful. Remember that leaving a brake shoe dragging will generate heat that causes brake fade and also accelerates shoe wear.

Adjusting a twin leading shoe arrangement is similar, but needs to be carried out twice as each shoe has its own adjuster. These can be at either end of the shoes, but are usually at the cylinder end and adjusted through a hole in the face of the drum. Some older cars have a hole in the wheel that lines up with the hole in the drum, allowing for rapid adjustment as this used to be a very regular task. Back off both adjusters, then adjust one shoe at a time. Only move on to the second shoe once the first has been correctly adjusted. Remember that adjustment is intended to take up the free play introduced due to wear of the friction materials, so pay attention to the wear of the shoes.

REMOVAL, INSPECTION AND ADJUSTMENT

The pictures in the following sequence show the removal of the rear drum, inspection of the system and readjustment after reassembly. The vehicle used is a 1971 VW Transporter. Remember that drum brakes fitted to older vehicles carry an increased risk of containing asbestos. Always take every precaution to minimize your exposure to asbestos as it is extremely damaging to health.

With the car safely located on four axle stands, one at each jacking point, the relevant road wheels will need to be removed. Before you do so, however, this is a good opportunity to spin the wheel you are about to remove, as the larger diameter provided by the tyre gives more leverage than trying to turn it by the drum alone. Where possible avoid turning the wheel with a bar on the wheel studs; this can cause damage that could result in the thread failing or, even worse, the stud cracking. Take off the handbrake; spinning the wheel once the handbrake has

been disengaged will allow the brake shoes to centre and make removing the drum easier. If there are any retaining screws these should now be removed. Treat them with care, and if damaged they should be replaced. In the event that they are too stiff to remove, an impact driver can be used but the screw should then be replaced. If the drum does not come off easily, a hide-faced mallet can be used on the strengthening rim around the outer edge, tapping at alternating sides in order to walk the drum off the shoes. If you have to resort to mechanical leverage to remove the drum, make sure you do not damage or distort the backplate. You may find it easier to disconnect the handbrake to more fully retract the mechanism. At this point an ice cream container or similar is ideal for storing any items you remove. If the system is unfamiliar it is also advisable to take photos of the spring and shoe arrangement to enable you to reassemble it, although that does make the assumption that it was assembled correctly to begin with.

The road wheel removed. Note the two locating bolts in the drum.

The drum removed. Removal is easier if the shoe adjusters are slackened off until they are no longer in contact with the inner face of the drum.

This picture clearly shows the single wheel cylinder in an assembly of one leading shoe and one trailing. This is a good opportunity to check for fluid leaks and that the dust caps are undamaged.

The handbrake cable hooks onto a lever on the rear shoe.

Shoes, springs and adjusters removed, showing the backplate and the handbrake cable.

With the drum removed, assess the condition of the drum itself, paying attention to the smoothness of the inner face and checking for cracks and damage. Mild corrosion can be smoothed down with abrasive paper, but the presence of corrosion on the inner surface of a drum that has been recently used raises questions about the operating efficiency of the rest of the system. If there is damage to the drum it should be replaced and fresh shoes installed. The shoes are normally retained with small springs. Depending on the design, the springs may be removed by simply unclipping them, rotating their slotted section by 90 degrees or sliding them off their locating peg. Be careful not to allow the spring to fire itself across the workshop at this point. The shoe retraction springs usually link the diametrically opposed shoes, pulling them back after the pressure from the slave cylinders has been released. These springs can be unhooked, allowing the shoes to be removed and assessed. The friction material should not be falling apart, considerably thinner in one spot, or below minimum wear specifications. Drum brakes typically wear more slowly than disc brakes due to the larger friction material surface area. Do not throw the old shoes away until fresh ones have been sourced; in the event that it is difficult to find replacements the old ones can be professionally relined by friction material specialists.

The remaining mechanism can be removed (with more photos taken to assist with reassembly if needed) and placed on a surface to allow it to be inspected for wear and damage. The backplate should be cleaned and inspected. The area directly in contact with the brake shoes may show minor wear due to the sliding action of the shoe. If this is excessive the backplate will need to be replaced, but this is seldom an issue. The slave cylinders can now be checked for wear, leaks and perishing of the rubbers. In the event that they are leaking or damaged, the simplest thing to do is replace them with an identical replacement; if this is not available a rebuild can be attempted using fresh rubbers.

The components can now be reassembled with new materials where required. The hardest part is probably the installation of the various springs, although special tools can be purchased that will help with this; they seem expensive until you discover how much easier the task is when you have the right equipment.

A tiny amount of copper grease can be used to lubricate the moving parts (with the exception of any friction materials or components that come into contact with them), but this should be kept to the minimum. Copper grease is designed to retain its lubricating properties in conditions of extreme heat, hence its use in braking systems. This can work both ways, however, since unlike most lubricants it will not burn off when the brakes are used, so contamination of the friction material is quite a serious matter. Excessive grease build-up will trap dust and make an abrasive paste that is not healthy for mechanical components.

When you install the adjusters, make sure you fully retract them first and that they move freely.

The component parts of the drum brake.

The inner surface of the drum should be kept clean to ensure that it mates accurately during installation.

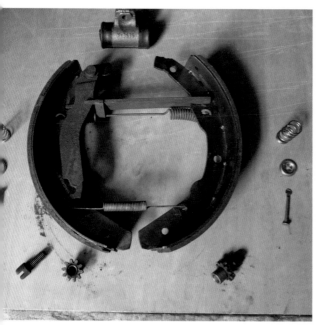

Inspect the surface of the drum. Small cracks can be an early indication of more serious failure in the future. This drum appears to be perfectly serviceable.

Shoe retaining spring.

In this application the star adjuster moves the shoe closer to the drum, removing free play.

Wheel cylinder assembly, showing the bleed nipple (dust cap on) and one piston removed.

The star adjuster nut fits on the slotted bolt for shoe adjustment.

The surface of the shoe.

Rear shoe, showing handbrake lever assembly and spacer bar.

Checking the shoe for friction material thickness and contamination from axle oil or brake fluid. A contaminated shoe should be replaced.

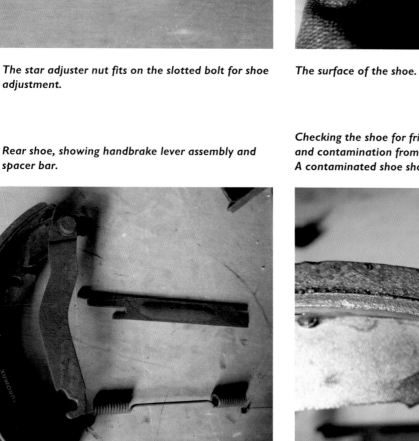

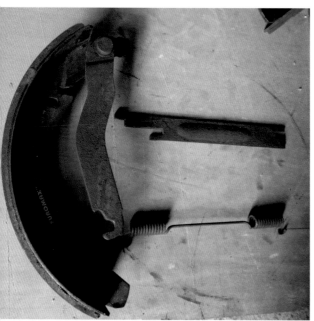

Backplate ready for reassembly. The wheel cylinder has a wire retaining the pistons to avoid accidental loss of brake fluid.

The adjuster fitted into position. The shoe sits in the slotted head and the star nut is turned by levering a screwdriver through the aperture in the backplate. A little copper grease on the thread and housing body helps prevent seizure and makes adjustment easier.

Housing for adjuster. Adjustment is carried out using a screwdriver through the hole in the backplate.

Adjuster and handbrake cable.

If they are of a screw head type, ensure that the head is undamaged and capable of being operated easily. In the event that the adjuster is automatic, its ratchet should be placed in its shortest position before reassembly.

Once all of the components have been refitted and the drum is in place, the system will need to be bled if the hydraulics were interrupted, and the brakes can be adjusted. With an automatic adjuster it is simply a matter of applying the brakes (with the handbrake cable installed, although this may vary depending on the installation); a ratcheting click is heard and on the second or third application of the brakes the shoes will have self-adjusted. With a normal mechanical adjuster it is best to replace the road wheel, assuming you still have access to the adjustment. If the adjustment is not accessible with the wheel on, then simply fit the wheel nuts and tighten them evenly without the wheel in place (using spacers if required to prevent them going thread-bound). The brake shoes can then be adjusted up to the drum to the point at which they drag the drum, and then retracted one click at a time until the drag no longer occurs. The brakes should then be applied and the adjustment checked again. Once this has been carried out the handbrake mechanism can be reconnected and the car returned to a drivable condition. It is quite likely another adjustment will need to be carried out after a few miles' driving as the shoes bed themselves in.

Shoe assembly, viewed from the front.

Shoes in position on the adjusters.

Shoe assembly, viewed from the rear. The handbrake cable can just be seen located at the base of the operating lever.

The backplate, showing the hydraulic pipe entering the wheel cylinder, with the bleed nipple above.

Assembly showing the wheel cylinder at top and adjusters at the bottom.

Holes in the backplate, with the star adjusters visible.

Adjust one brake at a time, turning until the shoe is hard against the drum, then slacken off a little until the drum rotates easily. This may need repeating after a few miles of use as the material compresses.

AUTOMATIC ADJUSTMENT

Modern drum brakes use an automatically applied adjuster, usually in the form of a ratchet that takes up any free play in the system. As the system wears and more play becomes apparent, the ratchet will adjust incrementally the amount of space it takes up, removing the slack. When components such as the shoes are replaced with new items, the ratchet needs to be reset to its initial position: on first application of the brakes after maintenance the ratchet will click to adjust out the residual play. Care must be taken not to operate the brake pedal or handbrake with the drum removed, as the ratchet will see the free movement of the shoes as free play and attempt to adjust itself to the maximum, resulting at best in the requirement to reset the mechanism, and at worst damage to the adjuster. The adjuster is often made of plastic and can be easily broken, so should be handled with care.

CHANGING THE FRICTION MATERIAL

On older shoes the friction material was held in place with metal rivets. Today, however, most shoes are bonded in place with a chemical process. In case you are struggling to find replacements for an older set of shoes, there are specialists who still have the required equipment to install fresh modern friction material to old shoes. This process, which used to be common practice at most local garages, usually involves cutting a length of friction material from a roll and punching it with holes that match the material removed. The material is then riveted on to the shoe and trimmed to shape. Remember that old shoes are quite likely to contain asbestos, a substance to which exposure should be minimized at all cost.

CHAPTER FOUR

HANDBRAKES

HANDBRAKE MECHANISMS IN DISC BRAKES

One of the biggest drawbacks with disc brakes is the difficulty in operating them as a handbrake with a cable mechanism. Although the handbrake is mainly used for holding the car stationary, when not in use its most important function is as a back-up for the primary braking system in the event of a brake failure. The handbrake usually operates on the rear wheels, and traditionally consists of cables operating levers that in turn push the brake shoes into the drum. With a pad system the technique is much harder to apply as the pads are smaller, which gives less room for an operating mechanism, and the relative force required to operate the brakes is quite high.

A few options are possible other than direct mechanical operation of the standard pads. The first is to use a small drum brake hidden within the centre

The handbrake drum hidden under a disc brake bell on a Vauxhall (disc not shown). MARK RICHARDSON

of the brake disc. If you are having difficulty finding your handbrake mechanism, there is a good chance it is hiding there. The advantages of this system are that it is usually quite effective, the shoes are not actually used for stopping the car and so don't suffer much wear, and the two systems are very much separate, so a total failure of one will not usually result in a failure of the other. The main disadvantage, of course, is the additional complexity and weight: two entire braking systems in one wheel is expensive and could be considered a cumbersome solution to the problem. This system is often used on larger luxury cars where the additional weight is a drop in the ocean, although it can also be seen on lower cost vehicles.

Another technique for applying a handbrake mechanism to a disc brake is to use an entirely separate mechanical caliper. This is usually reserved for the brakes of expensive performance cars, which use large calipers intended for motor sport or high speed use. Thus a separate caliper is installed on a different section of the disc to allow a back-up brake system to be operated. These systems are relatively simple to care for, being essentially just a mechanically operated caliper attached to a handbrake cable. Maintenance is usually nothing more than inspection to check for correct operation and pad wear.

REBUILDING HANDBRAKE MECHANISMS WITHIN THE CALIPER

The handbrake introduces a mechanical complication into an otherwise relatively simple hydraulic system. A disc brake with a handbrake incorporated is constructed usually in one of three ways:

- A lever system operating the caliper and pads directly

A handbrake mechanism on a brake caliper. The mechanism is more exposed to the elements than it would be on a drum brake, and has to exert more force.

- A separate smaller caliper mounted somewhere else on the disc entirely separate from the main braking caliper
- A drum brake integrated into the hub of the disc brake

The most desirable of these is probably the lever system, as it is the simplest and means the unsprung weight is kept low. The integrated drum brake is probably the most effective system as drum brakes tend to provide better hold-on pressure than a caliper-based system. Drum or separate caliper systems should have almost zero wear as the friction material is only used to hold the car once it is already stationary or during emergency braking in the event of a hydraulic system failure.

Stripping these systems is best carried out using a brake cleaner solvent. Reassembly should be carried out using copper grease wherever there are components that will slide over each other. The handbrake mechanism is often neglected and seized, especially on automatic vehicles that may be left with their transmission in Park without the handbrake being engaged, since the locked gearbox prevents the car from rolling. As a result the handbrake system could go for weeks or months without actually being operated. A stiff handbrake mechanism can cause the friction material to drag on the disc or drum, leading to overheating and unbalanced pad wear.

Stripping the handbrake mechanism is usually carried out in parallel with dismantling the rest of the caliper. A pencil and paper to make notes of the assembly, or a camera to take strategic photographs, will be a huge benefit here. Use brake cleaner to remove all grease and deposits from the system. Don't forget that an excess of even clean grease can encourage the build-up of abrasive silt from the pads, so use any grease sparingly. Even though the handbrake mechanism should not see the same levels of heat as the rest of the caliper, it is still worth using a copper-based grease just to be certain.

As the handbrake system is only operated when the car is stationary, except in an emergency, it does not really get subjected to application vibration, friction heat or many other stresses. This can actually be a disadvantage as it can mean the system is slightly more likely to seize due to a build-up of moisture and corrosion. A good practice is to apply gentle handbrake pressure occasionally, perhaps once or twice a month, to stop or slow the car from low speed. When stopping at traffic lights, for example, the initial deceleration could be carried out in the normal way, but the last few feet from a single figure speed could be arrested using gentle application of the handbrake. This removes corrosion from the inside of the drum, gives the handbrake system a functional test and exercises the whole mechanism. It sometimes happens that the handbrake on an automatic car is only operated once a year for the annual MoT test. For this reason it is especially important that you take the opportunity to exercise the handbrake safely once or twice a month. It is important to be aware of some important safety notes:

- The handbrake usually operates on the rear wheels only. These brakes are not as effective as using all four brakes. As a result you should leave plenty of stopping distance, especially if you are unsure of how much stopping the car will need.

- The handbrake is a mechanical override of the normal braking system, and as a result the ABS system *will not* stop the wheels from skidding. Apply the handbrake gently and, if the wheel locks, release it and reapply, or apply the main brakes as normal.

- Operating just the rear brakes could result in an unintentional handbrake turn. Use the handbrake gently and do not sharply apply it or the wheels may lock up.

- Don't test your handbrake with anything in front of the car. Choose a safe area where there are no safety implications if the car does not come to a halt as quickly as intended.

HANDBRAKE OPERATION

Of necessity a handbrake should be mechanically simple so that it operates with little fuss in an emergency. When you pull upwards on the handbrake lever, it pivots from a fixed fulcrum at the rear of the lever and applies tension to a Bowden cable running to the rear wheels. A sprung ratchet means that when the lever is released it does not drop back down releasing the brake again, so the vehicle can be held with the brakes on. The cable running to the brakes usually has to be split to run to each rear wheel, either at the handbrake mechanism itself, where one cable could be attached at each side, or through the use of an equalizing mechanism. This mechanism could be as simple as a pulley or curved plate through which is run a single cable linking the rear wheels.

When the single cable pulls the pulley the slack is taken up on each side evenly by allowing the cable linking the rear wheels to centralize over the pulley. Once the tension is even on each side of the pulley, equal force is applied to both rear brake mechanisms. When the button on the handbrake is depressed the lever can be allowed to rest back in its lowered position, the tension on the cable is released and the brake pressure relaxed. Where there is excess travel in the mechanism, due to wear in the friction materials or stretching of the operating cables, there are usually a couple of places that adjustment can take place.

It is a common misconception that MoT testers are looking for a specific number of ratchet clicks when the handbrake is applied; in fact they are looking for the brake to operate with a sensible amount of reserve travel left over. If the brake is almost fully applied but the lever has run out of authority then obviously the system is inadequate. Adjustments at the handbrake end of the cable, such as turnbuckles or threaded adjusters on the cable mounting points, are usually intended for removing wasted motion from the cable and not the friction materials. Adjustments closer to the brakes themselves are usually to adjust out friction material wear, and should be adjusted first. If the cable is still applying the brakes when the lever is down, it should be loosened at the handbrake lever end before trying to make any adjustments at the brakes. Once the cable has been confirmed as not having an effect on

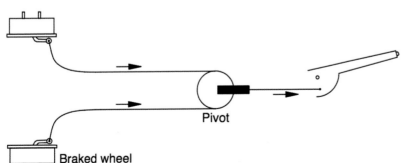

A handbrake using a pulley to equalize the force applied to the two rear brakes.

the brake application, adjustment can take place. In the case of drums the adjustment is usually the same as that detailed for the shoes (*see* Chapter 3). Sometimes an automatic ratchet adjustment is provided within the drum to take out some of the cable slack. This mechanism needs to be studied and reset to the beginning of its motion, especially if the drum or shoes have been changed. After this mechanism has been reset, the first application of the handbrake, which should only occur once the system has been fully reassembled and the drum refitted, is likely to travel a long way before operating the brakes; a ratcheting noise may also be heard as the system takes out the lost motion. On subsequent operations of the lever the brake should operate normally, with the correct amount of motion.

Some rear disc brakes have a small internal adjustment to take up the space between the handbrake operating mechanism and the piston. When the pads are changed this adjustment should be wound back to allow the pistons to be retracted. This adjuster can sometimes be well hidden; an Allen head adjuster could be hidden under a bolt head, a slotted screw head under a rubber bung, the variations are seemingly endless. Turn the adjuster with care, and without too much force, as the adjustment mechanism can sometimes be weak and threads easily stripped. If possible obtain diagrams and instructions for the caliper to ensure you move the adjuster in the correct direction. For a typical pad change the pads would be fitted and the brakes applied via the pedal for a few applications before adjusting the handbrake actuator. The adjuster would usually be tightened until the pads start to drag and then turned back a small distance, about a quarter of a turn. Only once this adjustment has taken place on both sides can the cable be adjusted to remove the free motion. It may need to be adjusted again once the pads have bedded in, and then maybe again as the pads wear further and the handbrake appears to have more travel than is reasonable.

The handbrake cables should be lubricated with care; some lubricants cause more harm than good by allowing dust and grit to stick to the cable. This, is turn, abrasively wears down the mechanism and cable guides. Most modern cables have a liner of Teflon or similar, which allows the cable to slide easily without attracting abrasive particles. These should not be lubricated, or lubricated sparingly with an oil designed to work on Teflon sleeves. The pivot points and pulleys can be lubricated with a light oil, although copper grease is preferred where the heat of the caliper can be transferred, such as levers on the caliper itself.

Some handbrakes, such as those found on Land Rover vehicles, may operate on the transmission directly. Great care should be taken when operating these brakes as you can damage the transmission. Apart from occasional very light application to clear the inside of the drum of moisture and/or corrosion, it is probably best to leave these brakes alone unless parking or in an emergency. In theory these brakes apply themselves to the driven wheels, but since they are applied before the differential, if one wheel is on a slippery surface it can spin, which in turn means no braking force is applied to the opposite wheel. Therefore even with the handbrake applied hard the car could still move. Treat transmission brakes with care.

FLY-OFF HANDBRAKES

The term fly-off handbrake can refer to a couple of variations on a theme. Some vehicles where the handbrake would interfere with entering or exiting the car when engaged were fitted with a handbrake that could be pushed down out of the way when not in use. Although the brake would remain applied, the lever could be dropped flush to the floor. This was usually specified on older sports cars where the handbrake lever was mounted between the seat and the door. An additional pivot was generally fitted on the handbrake lever, allowing the handbrake to be folded down out of the way, but applying force to the mechanism once it had been extended again.

The other form of fly-off handbrake refers to a modification to the mechanism to allow the handbrake to be operated without the ratchet engaging. This means that the lever will return to its lowered position when released, releasing the brake immediately. This is primarily of use on competition vehicles and is simpler than installing a hydraulic handbrake. If the ratchet needs to be engaged for parking purposes, the button on the end of the lever, which on normal systems would *release* the ratchet mechanism, is depressed and the pawl pushed into the teeth of the ratchet. To release the ratchet, the handbrake lever is pulled slightly upwards and the lightly sprung button usually pops out, allowing the handbrake to be released. To modify a normal handbrake to a fly-off system is usually a relatively simple matter of moving the pivot point of the ratchet pawl so that the spring lightly holds it *off* rather than *on*, and the button pushes it into the teeth of the mechanism rather than disengaging it. Proceed with caution however: a fly-off handbrake is much easier to release accidentally than a standard handbrake and could be knocked off by a child or animal left in the car. Some MoT testers will fail a car fitted with a fly-off mechanism, as the MoT test point can be interpreted in a couple of ways. For cars that were fitted with this system as standard, such as the Aston Martin DB5 and some Triumph TRs, there should not be a problem as it was the production standard.

ELECTRIC PARKING BRAKES

Gadgets sell cars, and a recent gadget to hit the showrooms is the electronic parking brake. First released by Renault on their Vel Satis, the electronic parking brake is either an electric motor operating the same cables that would be operated on a more traditional system by the handbrake lever, or a motor mounted within the caliper operating the brakes directly. It is a godsend for those with issues with their mobility or the strength of their limbs, but for most it is largely no real improvement over the simpler handbrake lever. When the electronic handbrake button is applied while the car is stationary, or moving very slowly, the parking brake is applied rapidly to hold the car as swiftly as possible. On some vehicles it is also released rapidly as the management system notices the driver trying to drive away, through gear selection, rpm increase or other means, thus automating the process.

If the brake button is pressed at a higher road speed, the vehicle assumes that the main brakes have failed and a progressive but moderately rapid stop is required. The parking brake, or more accurately in this scenario the emergency brake, is gently and progressively applied bringing the vehicle to a controlled stop without locking the wheels: the driver usually needs to keep their finger on the button to progressively increase the braking force. The system will normally warn the driver that the brakes are being applied by illuminating a warning light and sounding an alarm to ensure the system has not been operated inadvertently. These braking systems usually self-apply when the key is removed from the ignition, which is fine if you are parking but not so useful if you want to work on the brakes. In this instance there is usually an override to prevent the brakes being applied. This could be something as simple as holding down the brake button while removing the ignition key, or pressing the pedal and button at the same time.

Remember that the system is automated, and theoretically could operate while you are working on it. Check the manufacturer's recommendations for isolating the system; if you cannot find any advice

One advantage of an electric parking brake is the reduced form factor in the cockpit. GREGORY MOOR

Integration with the vehicles on-board electrical system can greatly reduce the chance of accidentally releasing or operating the handbrake.

INSTALLING A HYDRAULIC HANDBRAKE

For rally cars, or those who want to go drifting, a hydraulic handbrake is a desirable piece of kit. Providing the ability to apply manually just the rear brakes, it allows the driver to lock the rear wheels and provoke a slide. The hydraulic handbrake usually consists of a master cylinder with a lever attached, either next to the handbrake or in a more vertical position so it can be rapidly grabbed from the steering wheel. When the lever is pulled the master cylinder pressurizes the rear brake line, applying the rear brakes. When it is released the pressure is released back into the system upstream of the handbrake. As the system is placed after the ABS unit (where fitted) it will lock the rear wheels, allowing a handbrake turn or drift to be initiated.

When installing a hydraulic handbrake you should mount the unit on a sturdy part of the car; mounting it on a flexible panel or where it risks coming loose is an invitation to an accident. The master cylinder is plumbed into existing brake lines by inserting your own pipework from the pedal master cylinder to the handbrake master cylinder, and in turn feeding this on to the rear brakes. On an existing system this is most simply carried out by cutting the line in a suitable place and inserting your own pipes, using an appropriate set of connectors. Some people prefer to use flexible pipelines to allow ease of installation; the neatest and most professional solution, however, is to install using Cunifer pipes, ensuring the pipelines are correctly supported with P-clamps along their entire length. If your braking system uses a split circuit with two brake lines going to the rear, then the only solutions available are to use a twin cylinder handbrake or to replumb the car from scratch. Some advocate using T-pieces to link the two brake lines, but this is very foolish as it removes the safety advantage given by a diagonally split braking system: dual redundancy. Keep the brake pipe systems separate, make all your connections to a high standard, mount the handbrake firmly, and bleed the system properly after you have installed it. It is important that you never rely on a hydraulic handbrake to hold the car

then it is prudent to disconnect the battery so the motor cannot operate. Of course this also means that if the car has a dead battery and the brake is engaged, you will not be able to release the brake electronically to move the car. As a result most vehicles with an electric parking brake have an override. On some vehicles, such as some Renault cars, the override is hidden under the boot floor and takes the form of an override handle, which is pulled to release the handbrake. Other marques, such as Subaru, provide a special tool in the emergency tool kit for releasing the brakes, which requires you to climb under the car. It goes without saying that, before operating any sort of override to release the brakes, you must ensure that the vehicle is firmly chocked to prevent it rolling away.

when parked. It is very difficult to maintain pressure on one of these systems since, as the brake calipers, pads and fluid cool, the brake pressure will reduce until the car is free to roll away. Fluid will seep past the seals in the master cylinder over time, resulting in a handbrake that no longer holds the car. Don't trust a hydraulic handbrake to hold your car when you are working on it, and always chock the wheels if you do not have a correctly functioning handbrake mechanism.

INSIDER TIP

While it is often claimed that a hydraulic hand-brake is illegal in the UK, you should remain within the law as long as you retain the mechanical handbrake as well. If in doubt, check with a local MoT centre. It is important to retain the mechanical system if only so you can stop the car if the hydraulic brakes fail. In America it is usually referred to as the emergency (or E) brake. This is far more descriptive of its intended purpose, especially as some cars, such as Mercedes, use foot-operated rather than hand-operated mechanical brakes in addition to the hydraulic foot brake.

CHAPTER FIVE

BRAKE COOLING

Friction braking systems convert the kinetic energy of forward motion into heat. The more efficiently this heat is dispersed, the more energy can be converted, meaning the efficiency of the brakes is improved. Brake cooling tends to be a consideration of the original design, with most systems using air as the main source of heat removal, although some competition systems have used water as an additional cooling medium. When a brake fades it is

This example shows the ducting facing the disc surface. Sending the cool air to the centre of the vented disc is much more effective, as it means both faces of the disc are cooled evenly from within as the air is displaced around the circumference of the disc.

usually because it has become saturated with heat, reducing the efficiency of the friction materials. One of the most straightforward ways to improve standard brakes therefore is to increase the amount of cool air getting to them. The shape of the disc during rotation effectively makes it act like a centrifugal fan, dragging air in through the centre and displacing it around the periphery of the disc via the cooling vanes. The cooler and more plentiful the air supplied, the better the disc will be able to shed heat, preventing or reducing fade and failure. Using a flexible air pipe designed for the purpose means you can collect cold air from the front of the car and duct it to the rear of the disc.

A suitable entry point for cold air can be created by removing fog lights, which are usually circular and about the right diameter for brake ducting hose. The hose can then be routed back to the suspension mounts, where it can be linked over to the wishbone. Remember to allow enough slack for the pipework to flex without becoming taut, and ensure you use hose specifically made for the purpose. The hose will be subjected to a large amount of airflow with the inherent buffeting, vibration, heat and moisture. It is worth paying for the correct spec. Ensure that when the wheel is on full lock it does not rub the hose, as even the expensive stuff is easily damaged by abrasion.

If a juddering is felt through the pedal or steering wheel under repeated braking (or any braking pattern that generates a great deal of heat in the disc), it is possible you have heated the disc to the extent where it has distorted. This is due to the metal sections of the brake expanding at different rates, which causes the surface to deform. Usually the problem can be sorted with additional cooling. The floating disc, however, has been designed to eliminate this problem altogether. Brake discs warp because the expansion of the disc is constricted at its centre owing to the shape of the bell at the centre of the rotor. If the disc is made of a separate piece of metal and connected to the hub by a series of locating buttons, the different expansion rates of the disc and hub are catered for. This is quite an expensive solution, but allows the disc friction area to be made of a hardwearing material, and the centre to be made of much lighter alloys that do not have to be resistant to wear. It prevents a great deal of the disc's heat from being transferred to the hub, and can also counteract minor misalignments between caliper and disc (although large misalignments should be avoided). Permanent disc warping is actually quite rare: a judder from cold usually indicates either an issue elsewhere, such as worn suspension bushes, or transfer of material from the pad to the disc. This transferred material builds up on the disc forming high spots, which generate the juddering sensation felt through the pedal and wheel. These can usually be removed with careful facing using a specialist disc grinding machine.

Some drum brakes are available with finning on the outer surface. The main advantage with this is the increased surface area, which gives improved heat radiation, and the fins themselves give additional rigidity to the drum. On some vintage race cars standard drums have been seen to burst under heavy braking, leading to a total loss of brakes on that wheel. As the cylinders then move much further, the effect is that the brake pedal effort is not sufficient to operate the brakes on the other wheels, which leads to a total loss of brakes on that hydraulic circuit. For this reason upgraded brake drums are certainly worthy of consideration. Some competition drums are made of aluminium, which negates the additional weight penalty that comes with fins; remember, however, that aluminium is a soft material and will not last anywhere near as long as cast iron. As a result of this most high quality aluminium drums use a cast iron liner to provide the friction surface. Combined with the excellent heat transfer capacity of aluminium, this gives excellent heat dissipation combined with reduced unsprung weight.

The trade-off is that aluminium drum brakes are almost certainly going to be more expensive than their cast iron equivalent. If drum brakes are your only source of stopping power, then it is fair to say that they make an excellent investment, even at a higher cost.

An important consideration regarding brake cooling is the construction of the road wheel. A steel wheel without ventilation holes, of a small size such as 13in, has a number of undesirable properties. Steel is not as good at conducting heat as aluminium alloy and therefore conducts less heat away from the disc. Lack of ventilation means a vented disc cannot flow air away from the centre of the hub as easily, and a smaller size gives less space for air to flow around the disc and caliper. In contrast, a larger aluminium alloy wheel with good spacing between its spokes conducts heat efficiently, allows good ventilation, and as a side benefit allows rapid inspection of the disc and pads without removing the wheel. Obviously there will be additional benefits, such as the potential to reduce the unsprung weight and use a larger tyre. Keep in mind that changing the offset of the wheel will affect the braking (see Chapter 15).

Some units are available that circulate the brake fluid on the application and release of the brakes. This means that every time the pedal is released the fluid returns from the caliper and fresh fluid is primed ready in the hose for the next application. Although this system sounds ideal as it should bleed itself, remove the need for easily damaged bleed nipples and keeps the brake fluid cooler, it is not without its problems. It is still necessary to change the brake fluid, and the procedure becomes more complicated due to the replacement of the bleed nipples with return lines. In addition, any contamination to the fluid is not held largely in one place, as it would be on a conventional setup, but is pumped around the whole hydraulic system. The additional pipework really sets this out as being more trouble than it is worth unless your budget, weight and bhp figures are substantial.

CHAPTER SIX

BRAKE BIAS

Master cylinders are available in different bore sizes, which in turn means a certain displacement of the pedal will displace a certain amount of fluid. If you use a caliper that is hungrier on fluid displacement than the original, you may find the brake pedal gains an unacceptable amount of travel. Don't make the mistake of thinking the travel gained is not enough to worry about: the amount a pedal moves when cold is very different from the pedal movement you might experience with hot brakes, hot fluid and

A pedal allowing some bias adjustment by the movement of the threaded rod at the base. In some applications this is operated remotely by a Bowden cable.

worn pads. Finding out you have made a mistake is best done on your driveway rather than at full pelt in top gear. Also to be considered with any upgrade of the brakes is the balance of the braking. Cars are naturally biased towards more powerful brakes at the front, as a car shifts its weight towards the front wheels during braking.

On a car without correct balance you can find that the rear wheels lock up first, effectively turning your braking manoeuvre into an unintended handbrake turn. If not enough braking power is directed to the rear wheels, then your brakes are not as efficient as they could be and could seize due to underuse. If you are uprating the front brakes it is prudent to also upgrade the rears, unless other upgrades negate these improvements. An example could be a front-wheel drive hatchback that has had the interior stripped. The reduction in weight over the rear axle, coupled with improved front brakes, could naturally cancel each other out. This is not an exact way to approach what is a complex physical science and should be approached with caution. Fitting a proper brake bias valve, adjustable in use, is often the most effective way of getting the most from your braking system.

Brake bias is usually achieved using one of the following techniques:

Correct bias by design: Careful consideration at the design stage, prior to manufacture, can ensure that the chambers of the master cylinder deliver the correct quantity of fluid at the right pressure. Almost all production cars will use this technique.

Brake pressure reduction valving: Valves designed to reduce the pressure of the fluid supplied to the braking circuit can take various forms. In motor sport a rotary or lever-operated valve can be used by the driver to change the bias. In many modern vehicles, especially commercial vehicles such as light vans, a brake proportioning valve is fitted to the rear axle. As heavier loads are placed in the vehicle, the valve is moved by the increased displacement of the suspension, which allows more braking pressure to be sent to the rear axle. In the event of the rear wheels 'going light', when the weight moving over the front wheels means less traction is available at the rear, the valve allows less braking effort to be sent to the rear, reducing the risk of lock-up. This has largely been replaced by effective ABS systems. Sticking brake proportioning valves can provide some interesting faults when they get stuck open or shut, especially on commercial vehicles that are seldom fully loaded.

Brake balance bars with twin master cylinders: Two master cylinders, sat side by side, have their input rods linked by a pivoted and threaded shaft. When the shaft is pressed directly in the centre by a rod connected to the pedal linkage, the pressure applied to both cylinders is exactly equal. A flexible cable attached to the threaded rod allows it to be rotated from the driver's seated position, allowing him to move the central pivot point by turning the threaded section. This biases the leverage one way or the other, resulting in a proportionally different pressure applied to each cylinder. This is probably the most effective method of manual brake biasing, and allows rapid and accurate adjustment during racing. Some small production-run sports cars have this system fitted as standard.

CHAPTER SEVEN

BRAKE LINES

The standard brake lines fitted to production cars are usually steel due to its low cost, strength and ease of forming. It is resistant to high pressures and is easy to manufacture. It does corrode, however, and it is a common MoT failure point on older vehicles. If a standard pipe is not available from the manufacturer, it is often more cost-effective to make your own from pipe available from most motor factors.

A neatly formed T-piece dividing the brake pressure to the rear wheels. GOLDSMITH AND YOUNG LTD

CUNIFER AND COPPER

Although malleable copper is more expensive to buy than steel pipe, it is much more suitable for bending in the home workshop and, of course, is very resistant to corrosion. Be aware that, although copper is flexible, it will work harden with every bend, so try to avoid bending it until you are ready to shape it to its final form. This is less of an issue with Cunifer, an alloy of copper, nickel and iron (the chemical names of which together form CuNiFer). Originally designed as a material suitable for use in pipelines that would resist the corrosive nature of seawater, it has found considerable use in the aftermarket automotive industry. Although still likely to work harden with multiple bends, it is more resistant to this than copper and maintains a more attractive finish over time. Resembling dull brass, it is otherwise very similar to copper to work but does not display the less desirable characteristics found in copper brake pipes. Wherever possible, Cunifer should be the pipeline material of choice. Cunifer is used as standard by some manufacturers, notably Volvo, who cite its corrosion resistance as their primary reasoning. It is sometimes known as Kunifer, but the trade-name Cunifer is the more correct spelling.

One disadvantage with both Cunifer and, especially, copper pipes is that they are likely to fracture if not correctly supported. As a result care must be taken to support the pipe using specialist plastic clips every ten to fifteen centimetres, or more frequently if there are lots of bends. Ensure the pipe is routed so that it does not chafe against any metalwork, and under no circumstances use it for supporting anything else. One or two individual cables zip-tied to it shouldn't do much harm, but hanging a wiring loom off of it is inviting disaster. Remember when routing the pipe that you may have to bleed the highest point in the system, so try to make the master cylinder the

A Pipe bending tool used to create neat 90 degree bends. This is useful to minimize the risk of crushing the pipe when being bent. JUSTIN WESTLEY

Creating a flare using a pipe flaring tool. Before starting, make sure the end of the pipe has been neatly cut and the tool faces are free of damage. JUSTIN WESTLEY

A pipe bent using the tool has a smooth radius and no sign of kinking. JUSTIN WESTLEY

A pipe bent by hand with visible constriction. Discard it and try again. JUSTIN WESTLEY

very highest point. When passing the pipe through bulkheads, you can either put a high quality grommet in the hole and pass the unflared pipe through, or you can make up a bulkhead connection. This usually consists of a union that bolts through the bulkhead and then connects each pipeline individually.

Although this technique potentially increases the risk of leaks by introducing another joint, it means the brake pipes can be made up in smaller sections, making them easier to repair if damaged. It also reduces the risk of vibration damage that could occur with a pipe going through a hole. The bulkhead connector is the recommended technique, despite the additional work it generates.

The process of flaring the end of the brake pipe to create a tight hydraulic seal is not easily learned. When purchasing brake pipes to lay into your car, buy a little more than is required to allow for mistakes and, preferably, to allow you to practise before making critical connections. Remember, if one of these connections leaks you will have a brake failure. If you have any doubt about the quality of your workmanship, discard the effort and try again. If you really can't get the hang of it, then its time to call in the professionals. This is no place for 'close enough' workmanship.

Where the brake pipe may be subjected to vibration or movement it is sensible to include a loop in the pipework. This gives increased flexibility to the rigid line without allowing too much vibration to be passed on to the couplings, as vibration can work harden the metal and cause cracking. When considering ease of installation, it also allows some leeway when installing the pipework to make fitting easier. A convenient method for putting regular sized loops in place, especially where there will be a few lines with a loop required, and therefore repeatability is important for neatness, is to bend the pipe around an aerosol can or similar. If smaller loops are required, a socket set provides some useful variety of sizes, but try not to make the loops too small as their effectiveness will be reduced. If more flex is required incorporate more loops, but try to keep the number to a minimum. As always you must be careful not to collapse the inner passageway as this will impede the fluid flow. Of course this loop has the potential to trap air, and therefore some thought should be given to how it is placed. Where possible, the loop should hang below the rest of the brake line, so that air is unlikely to get contained there. It should not cause a major problem if you cannot make the loop in this manner, since the force and flow generated during bleeding is usually enough to push bubbles out from any of the many high spots located throughout the braking system. Most vehicles use 3/16in brake pipe as standard, usually denoted as 4.76mm in metric. There are some exceptions, such as certain British

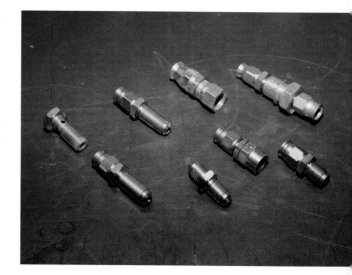

A selection of brake fittings. Consistency with union sizing is imperative to avoid the risk of mismatching components, which could later lead to failure.

cars that used 1.4in outside diameter tubing. The only sensible option is to measure what you have with vernier calipers. Bear in mind that layers of paint, grease, underseal or corrosion may have an effect on the diameter you read. Try to find a section of pipe in the best possible condition to allow your measurement to be as accurate as it can be.

CREATING FLARED ENDS

The type of flare you intend to use is important as a couple of variants are available. Some types flare the brake pipe internally to a Y-shape in cross-section (concave), whereas others create what is known as a bubble flare, which is more convex. Both have their advocates. I would suggest that you have a look at what is currently on the car and then carry on this trend. On many cars both types are used depending on the type of connection, for example from flexible coupling to fixed pipes, or fixed pipe to fixed pipe at a bulkhead. If you are making a system from scratch your choice is largely going to depend on availability. See what is available from your motor factors, and where possible duplicate what would have been on the car as standard. The two types of flare are not compatible and will not seal together, so make sure you get this right and stick to the same convention throughout. DIN, SAE, metric and US/Imperial sizes are available, so be very careful to ascertain the correct size and flare type. If there really are no other reasons to choose one type over the other, I would suggest making up the convex type, as these are slightly easier to manufacture using DIY flaring tools; however you may have differing experiences. Some flaring tools claim that they are able to make a 'universal flare' suitable for both the DIN and SAE sizes. While it is quite likely that this will be fit for purpose, it is sensible to obtain exactly the right flare for the fitting you intend to use. A compromise will always be just that.

A couple of tools are essential for making good flared ends on the copper pipe. The first is a good quality pipe cutter suitable for small-diameter copper tubing. The types utilizing a roller and rotary blade are perfect for this as they create a neat, symmetrical cut. The disadvantage with this cutter is that you will be making a bevelled cut; the sharp edge can be removed with a penknife or similar. When using this type of cutter, do not apply too much pressure with the thumbwheel as the cut will be irregular; smaller bites mean a more even and cleaner cut. It is possible to make a clean cut using a small hacksaw or Dremel type miniature cutting tool, but care must be take to keep the cut very accurate and without damaging the mating face. Any grazes or nicks taken out of the end of the pipe can easily turn into a split once the flaring tool is used. The tool of choice is most definitely the rotary cutter.

The other essential tool is a flaring device, which is intended to widen the end of the pipe in order to retain the pipe within the end of the nut. The inner surface of the pipe is flared out to come directly in contact with the male connector to provide a tight seal. As a result the flaring tool should provide a uniform and perfectly circular flare to ensure the mating has the highest possible chance of being successful. There are a couple of different designs on the market. The cheapest and most common consists of a die into which the pipe is clamped. Another section of the tool is then clamped to the die and turning a

wing nut or lever forces a cone-shaped former into the end of the pipe. This in turn forces the pipe out in the shape of the die. These are more difficult to use than the professional roller types, but are a fraction of the cost. There are a couple of ways you can make the process easier. A drip of brake fluid on the end of the pipe will act as a lubricant and will not contaminate the system as other lubricants might. Ensure the cut end of the pipe is as perfectly formed as possible before flaring. If you are still struggling, then try using a little more or less pipe poking from the top of the flaring tool. Of course, the most important advice when manufacturing brake lines is to put the union onto the pipe the correct way round, before starting the flare. The most perfect flaring will only be achieved seconds before you realize that the union you should have fitted first is staring at you from the workbench. If the other end is already flared, then it is time to get the cutter out again.

INSIDER TIP

The process of cutting the pipe usually starts to work harden the copper or Cunifer. If you are struggling to make a good flare, try filing back the end of the pipe with a fine file to make a surface that is perfectly flat. This fresh metal will be softer and more malleable than the work-hardened metal. Remember to remove any residual metal filings before installing the pipe. Don't forget the dribble of brake fluid or brake grease before making the flare; it has a big effect on the quality of the flare and the ease of manufacture.

If you are making a pipe in sections and need to join two pipes, it is better to use a double-ended male adaptor to join two female unions.

FLEXIBLE HOSES

The flexible hose is often replaced with a over-braided item in the belief that this will reduce pipe swell and that more of the motive effort applied by the pedal will be transferred to the piston. If your flexible couplings are swelling that badly then it is probably time to replace the pipes; even new standard items will give a noticeable improvement in pedal feel. The advantages of stainless steel over-braided hoses include better abrasion resistance, which may be of most use in harsh environments such as rallying, and, of course, the reduced swelling under pressure. Their disadvantages include the potential for the braid to corrode unless high quality stainless steel is used, slightly increased weight, and increased difficulty of inspection as the braid covers the pipework.

This flexible pipe has been routed so that it doesn't touch or rub any components when the wheel is on full lock in either direction, regardless of the suspension's varying state of compression.

ADAM GALLAWAY

CHAPTER EIGHT

BRAKE PEDALS AND PEDAL BOXES

The mounting of the brake pedal can be affected by a number of factors that are not immediately obvious. The rigidity of the bulkhead, especially on cars without servos where the brake pedal effort is greater, can affect just how much pedal effort is wasted in deflecting the sheet of metal to which the master cylinder and pedal assembly is attached. A good example of this is the Mazda MX5. The point where the pedal box and master cylinder/brake servo are attached the bulkhead is relatively

Although a neat solution inside the footwell, floor-mounted pedals can mean that access to the mechanics of the system becomes difficult.

GOLDSMITH AND YOUNG LTD

unbraced. As a result, when the pedal is pushed a good few millimetres of movement are wasted in the deflection of the panel. This might not sound like much but, when you are modulating the pedal to get the maximum force without lockup you may want to move the pedal millimetres back and forth to find the sweet spot. This bulkhead flex reduces feel and increases the sponginess. A common modification to the MX5 is to place a small bracket on the top of the driver's side damper turret. This is braced against the rear of the master cylinder, increasing the amount of force required to move the bulkhead. The result is improved feel and the sensation of increased braking efficiency. A neater solution is to ensure the bulkhead is braced correctly with reinforcement around the area most likely to flex. To discover where this is, ask a helper to operate the pedal (preferably with servo vacuum depleted) and watch the master cylinder and its mountings. You should be able to see any excessive movement and make up a bracing plate or triangulate to brackets around it accordingly.

These days it is uncommon to see a badly designed pedal box. They take a great deal of force from the driver's legs and need to be rigid yet light to operate. A good working knowledge of geometry is required to design a successful pedal box, ensuring that the leverage force created by the driver's foot is applied in a manner that serves as a compromise between the reasonable amount of effort that any driver may apply and the required distance of travel to operate the brakes correctly. To a certain extent this is also dependent on the ratio of fluid displaced by the master cylinder, something that needs to be considered during the design phase. Pedal spacing and length, and the space available are all limiting factors. If you choose to modify your standard pedal box you must ensure that your modifications do not compromise the strength of the unit in any way. If you insist on drilling or lightening components you must test them fully before use in earnest to ensure they do not fracture or snap.

Probably the most common modification pedal boxes receive in motor sport is to add an additional master cylinder to the brake pedal to allow the use of adjustable brake bias. Where the pedal originally operated directly upon the master cylinder, it now operates a threaded rod at 90 degrees to the direction of travel, which operates two separate cylinders via threaded turn buckles. Rotating the threaded rod changes the pivot point on the bar so that a deliberate geometric imbalance is introduced. This imbalance means that for a given amount of force and travel there is a proportional split in effort applied to the individual master cylinders, thus giving adjustable bias to the front and rear axle sets. Some pedal boxes are floor mounted, giving a more natural geometry to the pedals. You might expect that having the pedal pivoting in the same plane as the ankle would naturally offer a large advantage, but this is seldom noticeable to the driver as a majority of braking effort is applied by the leg itself, and not by the muscles of the foot. The design is commonly fitted in single-seaters, where keeping the height of the pedal box to a minimum to maintain a low silhouette is of prime importance to aerodynamic drag. By placing the cylinder directly behind the pedal, the amount of metalwork is reduced, the pedal box is kept simple and adjustments are relatively easy to work out. Adjustments in the rod between the cylinder and the pedal are usually carried out by varying the length of a threaded rod between the two. Most production cars had a solid bar here rather than a threaded bar, offering no adjustment. The pivot point of most pedals is usually a plain bearing. During restoration this can be stripped, cleaned and greased, but for the most part it normally requires little to no maintenance throughout the life of the car. If play is detected in the pedal, it is usually a good place to look first and damage here should be taken very seriously. If in doubt, replace all worn components with new.

If you are designing a pedal arrangement to suit yourself it is possible to have a certain amount of control over how much leverage is applied to the

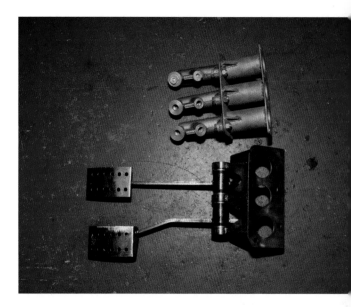

An incomplete pedal box part way through manufacture. Note the reinforcement and high quality of the welds. No compromise is acceptable when it comes to the strength of this component.
GOLDSMITH AND YOUNG LTD

master cylinder. Remember that the mechanical advantage gained by moving the fulcrum point and/ or attachment point of the master cylinder rod will result in increased travel. Before you start it pays to work out as much as possible about the amount of movement you will need and will be comfortable to drive with. It is incredibly hard to work out exactly how far the pedal will move, however, as there are so many variables that dictate the movement of the pedals, including the length and number of flexible lines, pad wear, the use of discs or drums (or a combination of the two), the number of brake lines used (for example, one circuit for the whole car, diagonal split, front and rear split), and the presence of hydraulic handbrakes or ABS. Annoyingly this makes the process rather trial and error. There are, however, some basic guidelines that you can use to assist you in selecting your components. Assuming you haven't deviated too far from the car's original braking system, you can select a brake caliper that closely matches the original brake caliper piston size. Calculate the original piston area using the formula

πr^2, and then multiply this figure by the number of slave pistons present or, if single piston sliding calipers are used, multiplying by two. This gives the chamber size per caliper, allowing you to select calipers with a similar displacement.

The system is not foolproof, but error can be minimized by using intelligent guesswork to predict whether your system will use more or less fluid flow and effort to operate the brakes. The first consideration is the use of discs or drums. An entirely disc brake system will require much less fluid displacement than a drum-based system, but with much increased force. Therefore it is reasonable to assume that a .750 sized master cylinder would be a good starting point for a drum-based system, whereas a .625 master cylinder might be a better starting point for an entirely disc-based system. This uses the premise that the .750 cylinder displaces more fluid but takes more effort, and therefore suits drums. The .625 cylinder displaces less fluid but requires less effort to move, suiting discs. However this is only a starting point for this example.

Real world applications have a lot more issues to consider. The goal is to get around 25 to 35mm of useful pedal travel from brakes fully off to brakes fully applied once free play is taken out. The displacement of the cylinder needs to be assessed and the pedal fulcrum and cylinder piston mounting then need to be trialled. If the pedal movement is too easy and the movement too long, the cylinder either has to be swapped out for a larger bore example or the mechanical advantage supplied by the pedal leverage needs to be adjusted. To decrease the travel, at the expense of increased effort, the master cylinder piston rod attachment point needs to be moved closer to the fulcrum point. If the effort is too high, the master cylinder piston rod attachment point should be moved further from the fulcrum point. This is usually a lot easier than lengthening the pedal, a dimension that is restricted by the need to maintain good ergonomics. It is of prime importance that the pedal never runs out of travel before the full required braking force is applied. Don't forget that the required pedal travel will vary depending on how worn the friction materials are, how old the brake

fluid is, the temperature of the fluid and working parts, and if there is any air in the system.

Always design in some contingency and ensure that your design never allows the master cylinder to exceed its intended travel in either direction. Doing so could either cause a catastrophic loss of brake fluid and pressure or damage the cylinder. Of course, if the system is working correctly it should never get to the stage where the cylinder exceeds its design limits, but there may be times such as bleeding or where a fault elsewhere in the system causes excessive travel. For this reason it makes sense to engineer in a stop, possibly along the lines of a bolt that can be locked in position or adjusted up or down to fine tune the pedal throw. One other point of adjustment is the rod between the pedal mechanism and the cylinder itself. On production cars this can be of fixed length or supplied threaded for adjustability. When it is adjustable you can adjust out the free play to a reasonable degree. Typically around 5mm to 20mm of free play is usual, depending on the car and set-up, but before you start making adjustments it is worth inspecting the system to see if any unusual changes have caused additional movement that wasn't there before. Before making any adjustments, the rest of the braking system needs to be totally serviceable to ensure you are not dialling out a fault elsewhere. Be aware that you should take care not to extend the adjustment too far as it may no longer be within the safe range of the threaded adjuster. Usually the adjuster uses lock nuts to prevent unwanted movement; if there is evidence that the adjuster has loosened itself, then it is prudent to use an additional form of locking, such as a thread-locking adhesive.

The intention is that the brake master cylinder is able to supply between 800 and 1200 psi to the brakes to allow them to operate efficiently. If your system is unable to supply this pressure then you need to reassess the pedal ratio and possibly the master cylinder size to find out why. Inserting a servo into the system should only really be done to make driving easier and more comfortable, not to compensate for a system that is too heavy to use normally due to poor design. If you are unable to get the required pressure without making the travel of the pedal too great, then consider the other variables in the system. For example, is there wasted movement in the mechanical parts of the system prior to the master cylinder? Free play there that has to be taken up before the hydraulics start to operate is counterproductive and can be indicative of serious wear. With higher pedal ratios, excessive travel due to free play is made more apparent by the increased movement: with a 3:1 pedal ratio, for example, the free play at the cylinder rod will be amplified by three times when it reaches the pedal, so 1mm of play becomes 3mm. Conversely a 7:1 pedal ratio would amplify any free play by seven, making the same master cylinder with the same 1mm of free travel generate 7mm of travel at the pedal. Of course, to a certain extent this is mitigated by the fact that a higher ratio pedal will require less total movement to generate the same pressure at the brakes, but all the same it is a factor that should be considered.

The great majority of the world drives on the right-hand side of the road and as a result most cars are designed with this as their preferred construction method. Where they are to be sold in markets that drive on the left, various modifications need to be made to move the controls over to that side. Most right-hand drive cars simply move the brake master cylinder over to the right to accommodate this, but in some applications this is not possible. In this instance some manufacturers use a linkage that takes the pedal motion from the pedal box across the car to the standard mounting position of the master cylinder. This is an inelegant solution, increasing the number of components required, weight, complexity and cost, but it is cheaper than re-engineering a large chunk of the vehicle to move the pedal box.

CHAPTER NINE

MASTER CYLINDERS AND BRAKE FLUID RESERVOIRS

When building a brake system from scratch or uprating an existing one you may need to consider choosing a different size master cylinder from standard. On the whole aftermarket master cylinder bore sizes are available in both imperial and metric measurements (with most popular being the imperial), with typical sizes being 0.625in, 0.70in and 0.75in. Usually in the imperial examples the 0 is dropped from before the decimal place, so you would typically order a .625in unit or similar. When going from single circuit to twin master cylinders it is usual to use a smaller cylinder on each, as you are displacing half the fluid you were before. With a bias pedal box small imbalances can be adjusted out, but for larger displacement issues you will normally need to change the size of the master cylinder to suit. In the event of too much travel being required to operate the brakes, a larger diameter cylinder will be used, but be careful to ensure the system is fully serviceable before deciding to make any changes. It could be that the excessive pedal travel is purely due to a fault in the system, such as air in the brake lines. Be careful not to use a cylinder that is too large as the amount of pedal pressure will increase. A fine balance needs to be struck between pedal travel and weight. Usually in systems running drum brakes on the rear a larger cylinder would be used for that circuit as more fluid is required to be displaced but less pedal effort required, for example .625in cylinder for the front axle and .70in for the rear. Of course this is not the only factor dictating the effort and pedal movement required (see Chapter 8). As a result the choice of master cylinder needs to take into account the type of pedal box being used and any adjustability available there.

Aftermarket master cylinders utilize either an integral or remote reservoir. The integral system is simpler, reducing plumbing headaches and providing a neat mounting solution. If space is an issue it is possible to use a remote system, but be sure to mount the reservoir higher than the master cylinder so that air bubbles can rise to the surface.

BRAKE FLUID RESERVOIRS

As the pads or shoes wear more fluid is required in the braking system to take up the additional movement of the pistons. For this reason braking systems are not usually completely sealed. They will naturally have a reservoir somewhere to allow extra fluid to be automatically introduced as and when required. Usually the reservoir is placed as high as possible, giving a natural point for air to be released, but there are times when this simply is not possible. Most cars integrate the brake fluid reservoir into or immediately onto the brake master cylinder, and this reservoir will often supply the clutch master cylinder as well to save on cost and complication. Standard production cars often integrate a sensor, usually a reed switch with a magnet mounted on a float, which will illuminate the brake warning lamp if the fluid level gets too low. This is useful firstly

There is very little room to manoeuvre in a Formula car. Filling and checking the reservoirs can require some ingenuity. GOLDSMITH AND YOUNG LTD

to give early indication of a leak or similar, or for giving early warning that the friction materials have worn enough to result in considerable fluid transfer from the reservoir into the main system. Either way the car should be stopped and further investigation carried out before going any further.

When the pedal is not being pressed the brake line is usually open to the reservoir, allowing pressure in the line to be released back to storage. This allows the brakes to be released and prevents pressure building up due to heat transfer or similar. On application of brake pressure from the pedal the master cylinder piston closes the brake lines from the reservoir and pressurizes the system. When the pedal is released again the line is reconnected to the reservoir and any additional pressure vented into storage once more. Some automatic bleeding devices use this characteristic to bleed the system without using the brake pedal by applying an artificial pressure to the brake reservoir. When each bleed nipple is opened in turn, the pressure in the reservoir forces fluid along the brake lines, although care must be taken to ensure the cylinder does not run dry as air will be introduced.

If you are installing your own pedal box with commonly available motor sport master cylinders, the most convenient option is to use the kind with an integral master cylinder, although these are usually part of the casting and it is difficult to check the level. Increasing the size of the cylinder reduces the flexibility of mounting places and it may be necessary to move the cylinder to a place that is inconvenient for checking and cleaning. Moving the reservoir to a separate unit has its own disadvantages: increased pipework means increased potential for leaks, although this system is only usually at atmospheric pressure, nowhere near the 1000 psi pressures experienced by the rest of the braking system. Also care has to be taken to mount the reservoir above the cylinder to allow air to be displaced into the highest point, namely the air gap at the top of the fluid where it can do least harm.

CHAPTER TEN

BRAKE LIGHTS

Operating the brake lights on a road car can be achieved in two ways: a mechanical switch on the pedal or a pressure switch in the hydraulic line. Where possible it is preferable to use a mechanical switch to reduce the number of fittings required in the hydraulic system. However, as hydraulic switches are available mounted in T-pieces it is sometimes possible to integrate them without modifying the system unduly. Mechanical switches are usually of the 'normally closed' type, in which current flows when not depressed.

The pedal makes contact with the switch when in its resting position, breaking the circuit to the brake lights. When depressed the pedal comes off the switch, usually mounted near the top of the pedal arm, closing the circuit and illuminating the brake lights. In the case of the hydraulic system the pressure switch closes when hydraulic pressure exceeds a set low value, sending current to the brake lights. Another technique for mounting a mechanical switch utilizes the rear brake switch from a motorcycle. Most brake lights operated by the rear brake on a bike use a microswitch connected to the pedal by a spring. When the brake is depressed the spring pulls the switch plunger, operating the lights. Any excessive movement past the limits of the switch is absorbed by the spring, preventing damage to the switch and ensuring the movement of the brake pedal is not restricted. Due to the flexibility of mounting possibilities and the variation in movement available, this system is useful when building kit cars or similar where a standard mounting switch arrangement is not ideal.

Other possible techniques include the use of a magnet and reed switch. When the brake pedal is not being depressed a magnet mounted on the pedal mechanism is held near a reed switch, which is held open by the magnetic field. When the pedal is depressed the magnet moves away, allowing the reed switch to shut and current to flow. Remember that if you choose to use a switch designed as a sensing device and not intended to carry the current of brake lights, then a relay should be used that employs the sensing action of the switch to flow a larger current without allowing the heavy current to pass through the switch itself. Using this technique a small and discreet switch can be used, with much thinner wires, allowing the relay to do the heavier switching.

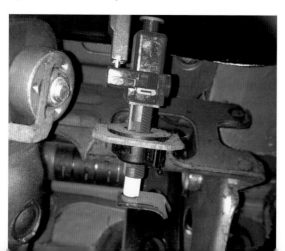

A brake light switch with the pedal at rest (brake lights not illuminated).

A brake light switch with the pedal depressed (brake lights illuminated). Note the threaded portion on the switch allowing for adjustment.

CHAPTER ELEVEN

BRAKE FLUIDS

A brake fluid tester establishing the moisture content, and therefore the boiling point, of the brake fluid in a vehicle. ATE/CONTINENTAL

By their very nature, hydraulic braking systems utilize a fluid medium to transfer the pressure applied at the pedal to the various slave cylinders around the car. Despite the manufacturers' recommendations, some cars may go for many years with the same brake fluid in the system that came from their initial construction, unchanged and uncomplaining. However, the fluid is possibly the most crucial part of the entire system and correct selection is too often overlooked as an essential part of the maintenance and upgrade process.

SAFETY

As a general rule brake fluid is damaging to paint, plastics and, most importantly, irritating to the skin and eyes. It is likely to be poisonous. Although the main constituent of brake fluid is often used in the manufacture of foodstuffs, there are likely to be plenty of other nasty chemicals in there that will do you harm. Read the warnings on the bottle. If you get any on your skin, rinse it off immediately. If you get any in your eyes, rinse with clean water until clear. If you suffer any ill effects, seek medical assistance straight away. Wearing comfortable goggles will mean you should never need to experience the pain of brake fluid in the eye.

Most people, when asked what is the most volatile fluid in their car, would say petrol. While petrol certainly gives off the most dangerous vapours,

An automated brake fluid replacement machine is the safest way of changing brake fluid as it removes the operator from a large part of the task. It does, of course, come at a price. ATE/CONTINENTAL

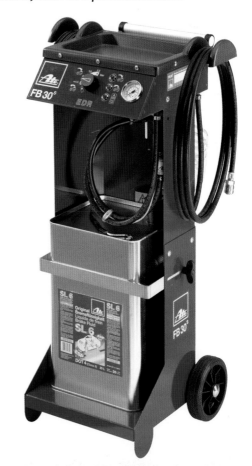

brake fluid is actually one of the most flammable fluids in a modern vehicle. It has a relatively low flashpoint, meaning it is more likely than petrol to burst into flame if it touches a hot component such as an exhaust manifold. Too many cars (and potentially lives) have been lost following a minor leak in a braking system that, when pressurized, sprayed onto components, causing the fluid to ignite. Treat brake fluid, and the connections in the braking system, with respect. If you have any doubts over the integrity of a component then spend the money and replace it. It is cheaper than replacing the vehicle.

THE DOT GRADING SYSTEM

DOT refers to the U.S. Department of Transport, one of the first organizations to suggest specific guidelines for the formulation and physical characteristics of brake fluid under a regulation called Federal Standard #116. This standard lists a number of criteria that the brake fluids must meet, covering everything from corrosion resistance to viscosity, but most importantly it states the thresholds at which the brake fluids will boil. As most brake fluids are hygroscopic, in that they absorb moisture, there is also a value stated for the 'wet' boiling point. This is the boiling point at which the brake fluid will boil when contaminated with 3.7 per cent water. When it has absorbed water the boiling point of the fluid drops considerably, increasing the risk of boiling in use, resulting in reduced braking force under hard driving. For this reason the importance of regular brake fluid changes can be understood, especially in competition vehicles where high brake temperatures are prevalent. As road cars are more likely to have older brake fluid, the wet boiling point is probably more significant. Race cars will have regular brake fluid changes, so the dry boiling point is more pertinent.

The most commonly seen brake fluid types are DOT 3, DOT 4, DOT 5 and DOT 5.1. The older DOT 1 and DOT 2 fluids are long obsolete; both had very low boiling points and were particularly hazardous to health. Where boiling points are stated, this should be considered the minimum boiling point required to meet the standard. DOT 3, DOT 4 and DOT 5.1 can be mixed together if needed, but will retain the standard of the lowest specification fluid present: DOT 3 mixed with DOT 4 will be considered equivalent to DOT 3, for example. DOT 5 cannot be mixed with any other fluid and it is extremely dangerous to try.

DOT 3

Still found surprisingly often, DOT3 has yet to be completely replaced in the marketplace by DOT 4. With a dry boiling point of 205°C it is a general use fluid, consisting primarily of a chemical called polyethylene glycol. It is becoming harder to find these days as it has the lowest specification and therefore is the least desirable fluid for use in modern applications. With a wet boiling point of 140°C, it is the worst affected of the DOT approved brake fluids when compromised by water contamination.

DOT 4

The most commonly used brake fluid, DOT 4 shares its polyethylene glycol base with DOT 3. It is formulated to have a higher boiling point, with dry boiling occurring at 230°C and wet boiling at 155°C. The majority of production cars are supplied with DOT 4 fluid, making it the most common 'standard' fluid.

DOT 5

Unusually, DOT 5 is a silicone-based fluid. This gives it much higher resistance to water ingress and a very stable viscosity. Whereas other fluids may show highly variable changes in thickness during temperature changes, DOT 5 remains fairly constant. It is also less likely to damage paintwork when spilled. It is not without its disadvantages, however: DOT 5 fluid can feel more spongy and gives less feel to the pedal, and is also prone to frothing. These characteristics make it unsuitable for use with ABS systems, which rely on the rapid movement of fluid in the lines to modulate the braking pressure. When used in a system that was previously filled with DOT 3 or DOT 4 fluid, a full and extensive flush is required to purge any trace of previous fluids. There is a high risk that seals previously used with other types of brake fluid may degrade at an accelerated rate when used with DOT 5. DOT standards state that DOT 5 fluid should boil at 260°C or higher, or 180°C when water is present.

Due to the many issues that DOT 5 fluid can present, it is seldom recommended except for an all-new braking system that does not use ABS. Because DOT 5 fluid mixes so badly with other fluids, it is usually coloured purple to reduce the risk of accidental mixing. Never under any circumstances top up DOT 5 with other brake fluids, or vice versa. Some vehicles (for example some Harley Davidson motorcycles) use DOT 5 from the factory. Do your research: if the system started off its life with DOT 5 then that is all you should ever use. The purple dye placed in DOT 5 fluid degrades over time, and provides a useful indication that fresh fluid is through to the caliper when bleeding the system. The old fluid will most likely come out the colour of honey, turning to a purple hue once the new fluid reaches the bleed nipple.

DOT 5.1

The limitations of DOT 5 instigated the formulation of DOT 5.1. Returning to a polyethylene glycol constitution, DOT 5.1 sets the highest standard for brake fluids in all characteristics tested by the DOT. With a dry boiling point of 270°C and a wet boiling point of 190°C, it has the highest resistance to temperature of any of the current standards. In systems where high temperatures are anticipated, DOT 5.1 is probably the most suitable choice. For a majority of systems it is overkill, with DOT 4 being a safe choice.

It is important to remember that the DOT standards are purely a standard that needs to be met; different brake fluids will meet or exceed that standard by different amounts. The constituent chemicals of brake fluids from different manufacturers are likely to vary depending on their own chemical mix; the DOT standard only states what the prime constituent should be. As a result, although in theory there should be no problem with mixing brands, it is best practice to stick to one brand at a time.

Once opened a container of brake fluid has a limited shelf life, especially where humidity is high.
ATE/CONTINENTAL

WATER INGRESS

The main limitation to brake fluid life, both on the shelf and in the vehicle, is the ingress of moisture. Water can creep in at unexpected places: a good example is the diffusion of water through the tiny pores in the flexible rubber hoses between the caliper and the car. Even when in good condition, these hoses are porous enough to allow water to wick into the fluid in tiny amounts. Over time this water can build up and cause issues. Of course, you may wonder why brake fluid is still so affected by water? When engine oils claim to be magnetic, and petrol is filled with detergents, why is it that brake fluid is still so crucially affected by the Achilles heel of hygroscopic attraction? The answer may surprise you: brake fluids are *engineered* to be hygroscopic. By attracting moisture and integrating it into their make-up, concentrations of water in the hydraulic lines are avoided. If a concentration of water is allowed to build up it can have a significantly lower boiling point than the fluid around it, causing it to flash boil into steam. In cold weather the water can turn to ice, blocking the brake pipe and preventing the brakes from operating. Crucially, when water is allowed to form inside the steel brake pipes, it can rapidly corrode the pipe from the inside out. This corrosion would not be noticed until the pipe fails structurally, causing an instantaneous brake failure just when they are being used hard. Suddenly the ability of brake fluid to absorb water looks like a very clever advantage rather than a difficult annoyance. Most water enters the braking system during top-ups and changes, through brake fluid that has been open to the atmosphere. For this reason, brake fluid should be purchased in the smallest containers you can and discarded after use. There should only be unopened bottles on your shelf. Brake fluid is not a consumable like engine oil; regular top-ups should not be needed so there is no need for a top-up bottle on the shelf. Brake fluid levels naturally drop as the pads wear, but return to full once the pads are replaced and the pistons displaced back to their starting positions.

SPECIAL FLUIDS

Certain cars use unusual types of brake fluid that should not be topped up or changed with normal DOT standard fluid. Due to the use of a hydraulic system shared with the suspension, Citroën cars use a fluid called LHM (or LHS on much older models). The fluid in this system is subject to very unusual environments that normal brake fluid would not survive, such as constant exposure to air and repeated pumping, which causes frothing. As more vehicles start to use increasingly complex hydraulics as part of their suspension and braking, the use of this fluid has spread to other marques, such as Rolls-Royce. The most recent standard of these fluids, called LDS, is further resistant to moisture ingress and has a longer service life.

CHAPTER TWELVE

BRAKE SERVO SYSTEMS

A freshly built servo ready to be installed in a classic car. Note the plated finish, which is preferable to a painted servo owing to its resistance to brake fluid.

GOLDSMITH AND YOUNG LTD

Drum brakes have a number of limitations, but one of their key advantages is the relatively low pedal force required to stop the car. In comparison, disc brake systems are heavy and usually require some assistance to make them acceptable for use in daily driving. A brake servo is a device that assists the driver in applying pressure to the hydraulic system operating the brakes. It can do this in a manner of ways, the most common being the vacuum servo.

VACUUM SERVO

The vacuum servo is best visualized as a large chamber with a diaphragm in the centre. This diaphragm seals one half from the other, but it is flexible so it can move when the air pressure differential between the two halves deflects it. The diaphragm is connected at its centre to the main piston that pressurizes the brake lines. Both sides of the chamber are maintained at a pressure well below ambient by means of connecting the servo to the inlet manifold via a one-way valve on petrol cars, or with an air pump on diesel vehicles. The air pump is required on diesel engines as they do not have a throttle body and so do not generate significant inlet manifold depression. When the brake pedal is

depressed, one side of the servo has its vacuum feed interrupted, and a valve is opened to atmosphere to allow ambient air pressure to fill it. This causes the diaphragm to deflect, operating the master cylinder. Once the diaphragm has deflected enough distance to match the travel of the pedal, the diaphragm is held in a state of equilibrium. If further pedal travel is applied, then the servo chambers become more imbalanced, which causes the diaphragm to deflect further. If the pedal is released a little, the atmospheric valve is shut and the manifold vacuum valve is opened until the shuttle moves back to match the pedal position, at which point the equilibrium state is held. If the pedal is released altogether, the vacuum is restored to the chamber and the diaphragm returns to its resting state. If the servo is in good working order the operating effect is such that very little pedal pressure is required to operate the brakes, since most of the effort is supplied by the vacuum from the inlet. The efficiency of a servo can be simply checked by repeatedly pressing the pedal firmly after stopping the engine. The pedal should move easily for the first few presses, then firm up. This shows that the servo had enough energy stored to carry out emergency assistance. Once the pedal has firmed up, maintain foot pressure and restart the engine. The pedal should smartly depress, showing that the servo assistance returns in good time. If the pedal assistance does not return there is an issue with the servo that will require diagnosis.

FAULTS

Although at first the vacuum brake servo may seem a complex unit, it actually has relatively few working parts. Problems usually occur due to vacuum leaks, either in the unit itself or in the vacuum pipework supplying it. This can sometimes be severe enough to cause running issues with the car as unmetered air enters the inlet manifold, causing the vehicle to run lean. If you have vacuum servo issues you should start by listening for air leaks. Be aware that the hose supplying the servo is often fitted with a one-way valve to allow the servo to retain its low air pressure during periods of wide open throttle (when the engine is not in a position to provide vacuum to the servo). If the leak is within the servo it will usually require a replacement unit as the internals are not user-serviceable, unless the servo is one of the older units for which rebuild kits are available. Other possible faults include a sticking input or output piston. This can happen for a number of reasons, but is usually due to old brake fluid carrying dissolved water, creating corrosion in the moving parts.

One of the more difficult servo faults to diagnose is the use of the wrong type of pipework supplying the vacuum. Most of the pipes on a car are designed to carry fluid under pressure, meaning that it can collapse when used as a way of supplying the servo with a vacuum. This often happens to the internal layers of the pipe only, with no visible change to the physical characteristics of the pipe. Make sure you use the correct pipe. You should always use a one-way valve to ensure that the servo retains its vacuum when the engine is running at wide-open throttle.

HYDRAULICALLY BOOSTED SERVOS

Some servos use hydraulic pressure to assist the driver in operating the brakes. This can be supplied in a number of ways. The first involves the use of a small electric pump directly connected to the master cylinder. This pump uses a pressure switch to maintain a small hydraulic accumulator at a pressure high enough to apply the brakes when the brake pedal is pushed, opening a servo valve in the process. This pressure is often in excess of 500 psi, with some systems operating as high as 2,000 psi. Due to these high pressures, extreme care must be taken when stripping these units: high pressure hydraulic fluid can cause terrible injuries. Always wear eye protection and do everything possible to ensure the system has depleted its pressure before removing it from the car. The simplest way to do this is to first disconnect the unit electrically, then pump the brake pedal until

the resistance increases. Pump a few times more and you are then safe to carry out the removal. If power is restored to the brake pump the process must be repeated. Do not apply power with any part of the unit stripped down as it may cause damage.

Systems used primarily on cars that have significant hydraulic systems already in place, such as Rolls-Royce and Citroën, utilize the pressure produced by the hydraulic pump to assist the braking. In the event of hydraulic pressure from these systems being lost, the braking will revert to its unassisted form, meaning the vehicle is still capable of being stopped but will require considerably more pedal effort than normal. Some simpler systems use a similar technique but utilize pressure from the power steering pump. The principles are the same.

When topping up hydraulically assisted systems ensure you use the correct fluid, which will differ depending on the manufacturer and system used. Do not make assumptions: some vehicles are very picky about the fluid used and mixing incompatible fluids can cause the system to fail. If in doubt, flush out the system and replace with the correct fluid. If you are fault finding a system that shares its hydraulic feed with other systems around the car, it is prudent to check if they are also working. Check that the fluid level is correct and then operate the different systems to ensure they function. This will allow you to trace the fault to either the pump side or the servo side.

IN-LINE SERVOS

On older cars that were not designed from the start to have a servo, or where certain under-bonnet constraints mean you cannot consider siting the servo in the usual place, an in-line servo can be used. In its simplest form an in-line servo is added to an existing system by cutting the main brake line and installing the unit in series. Whereas before the hydraulic pressure supplied by the master cylinder would be sent off to all of the individual slave cylinders around the car, an in-line servo puts a slave cylinder at the input of the servo. Thus the servo is operated when pedal pressure is applied, which in turn operates the hydraulics on the rest of the system.

CHAPTER THIRTEEN

ANTI-LOCK BRAKING SYSTEMS

HISTORY

Perceived by many to be a recent invention, anti-lock braking systems (ABS) have actually been around for many years. The first widely used, and purely mechanical, ABS systems came about in the early 1950s, most famously with the Dunlop Maxaret, but the actual concept predated the introduction of this unit. In 1908, for example, J. E. Francis introduced his 'Slip Prevention Regulator for Rail Vehicles'. In 1920 Gabriel Voisin experimented with a system that modulated the hydraulic braking system pressure on his aircraft brakes to reduce the risk of tyre slippage, something the high pressure and low contact patch aircraft tyres suffered from terribly. The first proper recognition of the ABS system came later with the German engineer Karl Wessel, whose system for modulating braking power was officially patented in 1928. Wessel, however, never produced a working product and the patent concept was as far as the system went. Eight years later Robert Bosch's engineering company produced a similar patent, but again the system did not go beyond the concept stage. Many technological advances were accelerated by the outbreak of the Second World War, but the production of ABS (an acronym that comes from the German term *Antiblockiersystem*) was largely marginalized.

The Maxaret system pioneered by Dunlop was used in initial tests on the CF-100 Canuck, a jet aircraft made by Avro Canada. The Royal Canadian Air Force had frequent problems with ice-covered runways. Tests using the Dunlop system showed a great deal of potential for reducing the risk of wheels locking during landing. Utilizing a spinning flywheel powered by a sprag clutch connected to the wheels, the Maxaret system allowed brake pressure to be applied as normal as long as the flywheel and the wheel speeds were relatively similar. When the wheel locked under braking, the sprag clutch would disengage, allowing the flywheel to keep spinning of its own accord. Once the flywheel had de-synchronized with the wheel unit to a certain angle, the brake pressure would be released, allowing the wheel to grip the road and spin back up to speed again. The braking pressure would then be automatically reapplied, and the process would start again if required. The system could operate ten times a second, much less than on modern systems, but still impressive for the time. The system proved itself on many aircraft types from the Hawker Hunter to the Handley Page Victor and English Electric Lightning.

The system was not applied to a road vehicle until 1958, when the Transport Research Laboratory, a British government-funded research organization, fitted the unit to the Royal Enfield Super Meteor, a relatively heavy 700cc motorcycle capable of 100mph, and happy to cruise at 80mph. The system proved to have remarkable safety benefits, including substantially reducing stopping distances on slippery surfaces, but it was deemed complex and expensive by Royal Enfield's management, who decided not to develop it further. The weight and expense, however, was less of an issue when fitted to cars. As with many motoring innovations, the Maxaret system was first used in anger in motor sport. The innovative Ferguson P99 from 1961 is perhaps best remembered for its incredible four wheel drive system, something that production cars were yet to utilize. Of more interest here, however, is that it was fitted with the Maxaret system. The P99 was extensively tested using the Dunlop system in place, but when the governing body introduced a rule change reducing the engine capacity by a considerable margin, the P99 was suddenly too heavy for its own good. Something had to go and as a result the P99 only ever raced without its Maxaret system fitted, managing a win at the Oulton Park Gold Cup. In 1966, however, Jensen fitted their powerful FF, a variant of the Interceptor utilizing a Ferguson four-wheel

The ABS ring mounted on the rear of the disc/hub.
MIKE SIMPSON

its comprehensive options list. Ford finally brought ABS to the European mainstream with its Scorpio in 1985. Made in Cologne, it was effectively a stretched Sierra with a few more luxuries, but it was almost certainly the standard fitment ABS that contributed to it winning Car of the Year in 1986, at a time when the motoring press was becoming more interested in the safety of a vehicle's occupants. Today there is not a single manufacturer that does not supply ABS as standard, even on the most basic models, as it has been mandatory in new cars sold in Europe since 2007. This fact, combined with improved tyre technology, undoubtedly saves thousands of lives a year.

OPERATION

Modern ABS systems are actually quite simple in operation once the initial shock at the apparent complexity has died away. The system can be broken into two halves: the sensing and the operating. Each wheel utilizes a toothed ring mounted at some point on the rotating components. This could be at either end of the driveshaft, but is usually on the wheel hub itself. Very close to this toothed ring is a Hall effect sensor, which produces a series of electrical

drive system, unheard of in cars of the time, with the Maxaret system. This made it one of the most technologically advanced cars of its era, but the Jensen was still very much a low volume production car.

ABS would not start to enter the mainstream until Chrysler introduced the 1971 Imperial. Some other American manufacturers, notably Lincoln and General Motors, had previously added rear-wheel only ABS to their cars, but it was the Imperial that first gave the world reliable four-wheel anti-skid technology. Japanese manufacturers followed shortly after when the Nissan President sported ABS on

An ABS ring on the inside of the hub. Visible at the 1 o'clock position is the ABS sensor, which detects the teeth on the ring. MARK RICHARDSON

In this close-up you can see the shape of the teeth that produce the waveform detected by the Hall effect sensor. MIKE SIMPSON

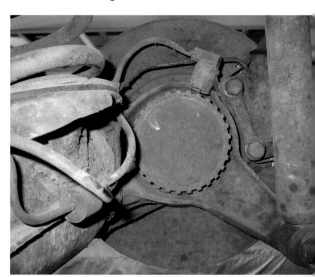

pulses as the teeth on the ring pass the sensing head. Normally a four-wheeled vehicle would have one sensing system per wheel, feeding back to a central ABS control unit. In normal operation the frequency of the pulses will be largely the same while driving. Sometimes slight variations may occur, such as while cornering, or if a tyre is slightly less inflated than the others (a phenomenon exploited by simple tyre pressure warning devices such as that fitted to the BMW Mini). Minor variations in speed are ignored by the ABS control unit, but if one wheel displays a sudden deceleration, signalled by a drop in the sensor frequency relative to the other wheels, the system knows that the wheel has locked up. If the frequency on one wheel *increases* then a traction control system, if fitted, may choose to reduce engine power or apply the brakes to that wheel only, as an increase in frequency indicates wheelspin.

Once a locked wheel has been detected the ABS control unit takes a number of actions. The hydraulic line to that wheel is blocked off from the pressure applied by the master cylinder by an electronically controlled valve. Some of the pressure is then released from the system between the brake caliper and the valve block, removing the braking force at that wheel. Once the wheel has resumed some speed, the pressure is then fed back into the brake line, reapplying the braking force. Should the wheel skid again, the cycle is then repeated. This may happen many times a second. The driver will feel the valves opening and closing as a pulsing through the brake pedal.

Modern ABS systems can be of a four-channel design, with each wheel having individual wheel sensors and actuation, or a three-channel system using four sensors but modulating the rear wheels as a pair. This system lacks some of the finesse of the four-channel system for a relatively minor reduction in cost. A variation on this system, primarily used on rear-wheel drive cars, uses individual sensors on the undriven wheels and a single sensor for the driven axle mounted on the differential. The relatively uncommon two-channel system, mainly used in American vehicles, has largely been phased out. This system would split the ABS into paired wheels and

was therefore less able to control individual wheel slip. The cost savings did not justify the reduction in capability and today most systems are three- or four-channel.

On vehicles without ABS it is common to fit a brake proportioning valve to modulate the pressure to the rear brakes, depending on the load over the axle. This is intended to give maximum brake pressure without locking up. Cars with ABS do not require this function, so the brake proportioning valve is usually omitted or simplified by removing the load sensing from it. This is fine while the ABS is active, but should it develop a fault or become immobilized (by removal of the fuse, for example), the system should not be assumed to be as effective as a normal vehicle without ABS. A car with disabled ABS is *less* effective than one that has not been fitted with ABS as there is no compensation to prevent rear-wheel lock-up. This could lock the rear wheels in the event of heavy braking with the associated loss of control. As a result you should not disable an ABS system without considering the correct regulation of brake pressure to the rear wheels.

ABS FAULTS

ABS is hydraulically pretty similar to a standard braking system and more often than not the faults that can manifest themselves on a standard system still apply. However, issues can arise with its very complex electronics, most commonly the wheel speed sensors, which can stop operating correctly. A wheel speed sensor is generally a Hall effect device comprising a metal ring, usually mounted on the driveshaft, with a number of castellations that pass very close to the ABS sensor. As the high points on the ring pass the sensor, they generate a small voltage in the coil of wiring within it, which is carried to the ABS unit. The number of pulses per revolution is directly proportional to the number of teeth on the ring, and therefore the ABS ECU can work out to a high degree of accuracy the speed of each wheel.

A number of different faults can prevent the sensor from doing its job, especially when you consider the environment in which it is operating. If the

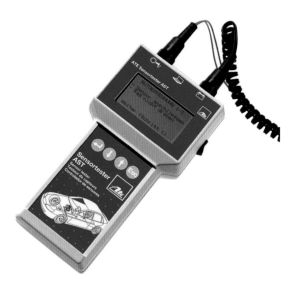

The ATE sensor tester is capable of interrogating the many sensors within the ABS system. Some of this data can also be obtained using standard garage diagnostic equipment via the EOBD/OBD-II port.

wire carrying the signal is damaged, the vehicle will no longer be able to see the pulses coming from that wheel and will consider itself to be in a fault mode. A warning lamp on the dash will illuminate and some systems will inhibit themselves. The same thing happens if the wiring inside the sensor fails. Modern ABS units are very sophisticated and are able to communicate this fault through means of a fault code, readable using a fault code reader, or sometimes through flashing the warning lamp in a set pattern when the fault display mode is entered, usually by bridging some terminals. The fault code reader will usually tell you specifically which sensor is at fault, removing the requirement for long and drawn-out fault finding.

Sometimes the electrical side of the system is in good working order and the fault is mechanical. As some manufacturers do not supply a new ABS sensor ring with their components, the ring has to be transferred from the old hub or driveshaft when it is replaced. This gives lots of scope for damage to the teeth, misalignment on reassembly, or even simply forgetting to refit the sensor ring. The ABS ring and sensor is usually placed on the hub right

next to the road wheel, meaning that all sorts of detritus can foul the ring and the air gap between ring and sensor. This will weaken the signal to the extent that an intermittent or solid fault can occur, especially if the dirt collecting on the system is metallic. Usually cleaning the dirt off completely is enough to get the system working again, at least until the next time it gets filthy. In the event of the fault lying with the sensor it should be replaced complete. Some vehicles additionally need the fault code to be cleared from the ECU using a fault code reader of the correct type. Cheaper fault code readers may only allow you to read the code and not clear it, so ensure you purchase the correct kind by checking with the manufacturer of the unit before you buy. The wire used to connect the ABS sensor to the ECU is usually coaxial, with the signal wire running down the centre of a cable that has a sheath of metal braiding to ensure the signal gets all the way to the ECU without interference or suppression from other sources. As a result, if there is damage to the ABS sensor cabling it is not usually an option to use any old wire you have in the toolbox; the correct loom should be installed.

Another fault that can cause the ABS lamp to illuminate is one that is actually used to good effect by certain manufacturers' flat tyre sensing systems. When a vehicle tyre loses pressure the weight of the car causes the tyre to deflect further than it would if the tyre pressure were normal. This means the effective rolling radius of the wheel is smaller, and therefore that wheel will turn slightly faster than the other three on a straight bit of road. Some ABS systems, such as that fitted to the BMW Mini, will alert the driver that one wheel is travelling at a speed disproportionate to that of the other wheels and should thus have its pressure checked. It is important to note that this system does not tell you that the tyre pressures are correct, only that one tyre's pressure is significantly different to that on the other side of the car. Be aware that the installation of this system should not take the place of normal and regular tyre pressure checks, as it is possible for an incorrect pressure to go unnoticed. On less sophisticated systems this pressure imbalance is either not detected or will cause the ABS system to suspect a fault, turning on the ABS warning lamp. Therefore, although it might seem strange, on seeing an ABS warning lamp illuminate the first thing to check is that all your tyre pressures are correct.

Failures of the ABS pump or valve block will usually completely inhibit the system and illuminate the warning lamp. The reason that even a small fault will disable the ABS system is due to the manufacturers' understandable caution about making the brakes do unpredictable or undesirable things when the driver is not expecting them. For example a faulty ABS wheel speed sensor will make the ABS system think that the wheel has become stationary, as if skidding. If the vehicle then releases the brake pressure from that brake line, the brakes are suddenly less effective as a wheel that has perfect traction will not be permitted to apply braking force. Therefore the ABS module carries out comprehensive self-check, hundreds of times a second, to be sure it can rely on all the different sensors it requires. In the case of the Hall effect sensors it can detect an open loop, caused by a broken wire or sensor, by passing a small current down the wire. If this current can no longer complete the circuit the ABS will assume that the sensor is at fault and illuminate the warning lamp. If it detects signals that are outside what the system considers normal, then again the ABS will be inhibited and the warning lamp illuminated. It is important to bear in mind that if the ABS is no longer working there will be no load proportioning to ensure that the rear wheels do not lock. Under heavy braking a severe instability could occur, leading to an accident. If the ABS pump or valve block develop a fault, such as sticking valves or blocked passageways, the ABS system will not be able to modulate braking pressure

or could even prevent brake pressure being applied to the brakes when required. Systems such as these are usually designed to be fail-safe, in that they will fail in such a way that they default to being able to operate, even if in a limited capacity.

Due to the additional complexity of ABS hydraulics some systems require an electronic box to be connected or a specific mode entered, typically by shorting some terminals on the connector, during brake bleeding to allow the valves to 'flutter' thus dispersing any air that has collected in the system further along the pipes to the bleed points where it can be bled off. Two undesirable things can happen if this mode isn't entered. The first is that certain valves within the ABS unit will remain shut, sealing old fluid in the system despite the rest of the fluid getting replaced. The other issue, slightly more serious, is that the complex shape of the valve chests will trap air, which is then unable to escape as critical escape routes are closed off. The process of entering the bleed mode causes the valves to flutter and agitate the fluid around them, encouraging air to be displaced and allowing all of the fluid within the system to be forced through by the bleeding process. Some ABS units have one or more bleed screws directly mounted on the unit itself; these are bled using exactly the same process one would use for bleeding the rest of the hydraulic system. In the event that your braking system requires the use of an electronic device to bleed it correctly, there will usually be a specific set of instructions to follow. Typically the system might run on a laptop or a hand-held stand-alone unit, and will give you instructions as you go. Typical of this are certain Volkswagen group vehicles, which use a system called VAG-com.

CHAPTER FOURTEEN

ELECTRIC BRAKING SYSTEMS

Some manufacturers have been experimenting with brake-by-wire electronic brakes for some years. Mercedes-Benz took the bold step of incorporating the Sensotronic system, co-developed by Daimler and Bosch and first used tentatively in the Mercedes SL class in 2001, into its 2003 Mercedes-Benz SLR McLaren. The Sensotronic system is capable of deciding exactly how much fluid and what pressure each caliper should be supplied with, meaning the risk of a wheel locking is greatly reduced and can be dealt with very rapidly should a skid occur. The presence of an electronic system between the driver and the brakes themselves makes it possible to refine the brakes, something that is increasingly difficult with high performance braking systems. Very powerful brakes, for example, may be 'grabby' and jerk the vehicle to a stop under braking with even minor pedal pressure; the Sensotronic system is able to modulate the pressure just before the car comes to a rest, making low speed stopping much smoother. The brake pedal itself is fitted with a simulated resistance that gives the driver the sensation of operating a normal braking system, simulating the amount of force required to stop the vehicle. Due to the complexity of this relatively new technology, Mercedes-Benz restricted their deployment of the system, retaining it for their highest performance models, although, as with many technologies found on high performance cars, over time it is likely to trickle down to the more everyday models on the market. Most of the benefits, such as the modulation of the braking pressure at low road speeds, are only really advantageous on the more aggressively braked models.

The system described above is a hybrid of electrical and standard hydraulic braking technology. There are two ways of incorporating true electric braking. The first is to use fully electric calipers, utilizing an electric motor to apply the friction materials to the disc. These calipers need to contain motors capable of operating very quickly, generating a great deal of torque, and capable of being in the presence of very high temperatures without degeneration. As with standard calipers, they are likely to get caked in dust, grit, mud and liberally soaked with water, sometimes when very hot, so they must be able to cope with thermal shock without detriment. As a result the technology tends to use components that are still quite expensive. A very desirable factor is the ability to work out exactly how much pressure has been applied to the disc via the pads. In theory it is relatively simple to embed strain gauges within the caliper to give accurate information on the deformation of the body of the caliper, allowing an accurate judgement of the applied force, but the brakes change temperature very rapidly, especially at the moment that braking pressure is applied. This makes it very difficult to judge applied pressure and a resolver is generally used to judge the position of the motor applying the force, and the braking force is calculated by incorporating the data provided by the resolver with the acceleration forces and other data.

The other form of electrical braking is more likely to be carried out on a hybrid or fully electric vehicle, using the motor that usually drives the car to provide a resistance to rotational motion without friction. This can be done by placing an electrical load over the motor, which can be a motor in the conventional position under the bonnet or somewhere in the body of the vehicle, or individual hub motors mounted in each wheel. This converts the energy into heat, or more usefully allows the motor to put charge back into the battery. One of the biggest advantages with this system is the complete absence of friction materials. As you are not using friction to slow the vehicle you do not require wearing parts (meaning less consumable components to service), and the energy, instead of being wasted to atmosphere as heat, can be sent back into storage for use during the next acceleration phase. Conventional brakes are usually used in conjunction with this type of braking, which when used to recharge the battery is known as regenerative braking.

The common factor with all the electrical systems mentioned above is the need for redundancy – additional sensors or systems used in parallel with the main system to take over in the event of the first system failing. Vehicles are fitted with a backup brake in the form of the handbrake, but this should only ever be considered as a last resort for stopping the car. On a modern production car every effort is made to ensure everything is reliable, from the air conditioning to the interior lighting, but the braking system undergoes far more extensive testing than any other system, especially where new technology is being exploited. In the case of electronic braking systems, the use of redundancy is critical. On hydraulically braked systems this might take the place of a dual circuit brake system, for example. On electrically braked systems some manufacturers might incorporate a master cylinder that is not normally pressurized under braking, but comes into use when the electronic braking system fails, meaning the electronic brake pedal defaults back into a crude hydraulic mode should the electronics fail. Whereas most modern cars use a single electronic switch on the brake pedal to switch on the brake lights and tell the ECU that the brakes have been applied, an electronic system will use a multitude of sensors on the pedal, not only so that rate and pressure of application can be judged accurately, but so that in the event of a single sensor failing there is still enough information supplied to the braking control unit to inform it that braking has been demanded. Some signals may get carried around the car on more than one data-bus, so that in the event of one being taken out of action there is still enough information being passed around the car to allow the brakes to operate correctly. Although these faults may illuminate warning lights and instruct the ECU to regulate the vehicle's top speed to allow it to be driven safely, allowance has been made for most faults that can reasonably be anticipated and compensation can be applied in the event of component failure.

CHAPTER FIFTEEN

SUSPENSION GEOMETRY

Changes to the geometry of a vehicle's suspension can have a drastic effect on braking. A simple change such as fitting wheels with increased offset can create increased loading on the steering when braking; if one wheel locks up, for example, the force transmitted through the steering on a car with high offset wheels, in which the centreline is placed further out from the centre of the steering action of the wheel, is much higher than that on a vehicle where the offset is such that the braking force acts evenly across the width of the tyre. Where the offset is zero, that is when the wheel pivots through the centre of its tread pattern, braking forces are largely balanced, which reduces the risk of steering snatch under heavy brake operation.

The toe angle of the wheels, which is the amount the front wheels splay outwards, will affect the stability of the car under braking. Many factors, however, can alter just how much toe it is desirable to set. For a typical mid-engined race car, some teams may set a few degrees of toe-out to give quicker turn-in when cornering at the expense of tyre wear. Remember that under braking the additional forces acting on

Setting the geometry. Experimenting here can yield improvements in stability and braking distances.
ERIC JACOBSEN

the braking wheels will change the geometry slightly, especially if running relatively soft rubber bushes. This means the toe-out will become more pronounced, or toe-in may get neutralized. This can vary depending on whether the steering rack is placed in front of or behind the pivoting point of the front wheels. For most cars a small amount of toe-in gives the most stable braking, but a car optimized for braking efficiency will be compromised elsewhere. As with most things, set-up is a careful balance of optimizing the geometry to cope well, whether braking, cornering or accelerating.

No matter how good the brakes are they will be unable to stop the car efficiently if the tyre is not fully in contact with the ground. For this reason the dynamics of the suspension and chassis should be set in such a manner that, when braking, the maximum contact area of the tyre is presented to the road. This is affected by the squat of the suspension and how the camber may change when braking loads are applied. Watch a large car with soft suspension when it is asked to brake hard: the suspension on the front almost collapses in height, and in turn the wheels often decrease their camber to the point where the inner edges of the tyre are presented to the road with more loading than the outers. On race cars the suspension is inherently a lot stiffer, reducing the dive and therefore reducing the change in camber under braking. This means that the optimum tyre contact is more likely to be maintained, assuming that the car has been set up correctly in the first place! Many of those lowering cars that are driven primarily on the road just fit a stiffer, lower suspension and do not consider the implications this may have for the geometry of the vehicle. A car with Macpherson strut suspension, for example, when lowered without any adjustment to the geometry, will present more negative camber than would normally be applied when stationary or moving at a steady speed. When cornering, body roll should cause the inner wheel to increase its negative camber if the lower arm is angled at less than 90 degrees to the Macpherson strut. If the vehicle has been lowered too far, the angle between strut and lower arm becomes more than a right angle, which in turn will force the wheel into gaining positive camber. In most circumstances this is undesirable for traction and will cause the car to become slower in corners due to the reduction in grip. Most cars, especially those designed for enthusiastic cornering, will be set up to maintain negative camber or to allow the camber to neutralize almost entirely on the wheel taking the majority of the cornering loads. Therefore when a car rolls in a corner the suspension geometry allows the loaded wheels to become almost totally upright with relation to the road, presenting the maximum tyre tread to the road surface in the most even pattern, while on the other side of the car the unloaded wheel should in theory increase its camber, negating the body roll that would otherwise compromise its contact with

the road surface. This will affect our braking needs surprisingly little, primarily due to the fact that nearly all the braking is done before the corner while the vehicle is not subjected to much in the way of sideways forces. Generally speaking, optimum braking is achieved when closer to neutral camber. Again, a compromise between the ideal camber for braking and cornering must be found.

When it comes to suspension damping rates the optimum rate needs to be found that allows the wheel to maintain contact with the road at all times. This is one of the few cases where we do not need to compromise. The best damping for cornering is often the ideal for good braking as both require the same thing: contact with the road surface. When suspension (and its damping) is stiffened, the weight transfer under braking pushes the front wheels of the car more firmly into the road surface, allowing more braking force to be applied before the front wheels lock up.

Note the daylight visible under the rear wheel of this car under braking. No matter how good your brakes are, they don't work on fresh air. SIMON CROSSE

CHAPTER SIXTEEN

THE ULTIMATE UPGRADE

As in competition driving, the most simple upgrade is often the one people are least likely to invest in. Braking the car on the road or track can be carried out more effectively after some simple driver training to ensure you are operating the system in the best possible manner. The braking technique will differ based on the car and engine layout, tyre type, road conditions, presence of ABS or other driver aids, and 101 other variables, but here are some guidelines that should at least assist with improving your technique.

FINDING THE POINT OF MAXIMUM RETARDATION

The point of maximum braking efficiency on a dry and grippy road is generally considered to be just before the wheels lock, when the wheels have started to move fractionally slower than the road beneath them. When braking hard with ABS this point is repeatedly passed through before the brakes are released and reapplied; as a result a car will actually stop more rapidly if this threshold is found and maintained until the vehicle is stationary. In wet weather this threshold is reached much more rapidly and is much more difficult to modulate, hence ABS gives a significant advantage. ABS makes no effort to find the 'sweet spot', but puts the brakes into a condition where they will, for a large proportion of the time, be operating in an efficient manner. The forces acting on the driver change infinitesimally, as the deceleration of the vehicle is modulated as the lock threshold is passed. An experienced competition driver is able to define this threshold and modulate the braking force without even realizing they are doing it.

If you are still learning this technique, it is a good idea to try out some heavy braking in a suitable and safe environment, training yourself to roll off the brake pedal pressure when the wheels lock. With practice you can learn to feel where the point of maximum grip lies and achieve maximum braking efficiency, no matter what the conditions. Sometimes even very skilled motor sport drivers will be seen to step the wrong side of the magic threshold, with one or more wheels locking up. As these drivers are usually right at the very edge of available grip, this may be enough to make them understeer and go wide at the corner they are trying to negotiate. Thus the importance of being able to find the perfect point of braking nirvana is constantly being practised and refined to gain the competitive edge.

Under braking the front wheels are suddenly asked to do more than just keep the front end of the car off the road. There is only a finite amount of traction available in a tyre, and if you are using maximum braking force there is none left for changing the course of the vehicle. Therefore in competition the most efficient way of braking before a corner is to use maximum braking effort until steering input is required, and then progressively roll off the braking force to free up some traction for the front wheels to negotiate the corner, a process known as trail braking. If you don't release enough braking force then the car will understeer; free up too much and you are not going to make the bend. It is the ability to make this judgement call automatically while under high pressure that is the difference between a racing driver and a wannabe.

MAINTAINING STEERING

One of the main advantages of ABS is that it removes the most dangerous problem of heavy braking: a locked wheel can have no steering input. Consider that the wheels are turned to negotiate a corner at an angle of around 15 degrees. While the wheels are capable of rolling they will modify the course of the car by an angle dictated by the steering input. As soon as those wheels lock they are effectively rubber blocks; a rubber brick in the shape of the tyre's contact patch, installed on the axle, would have much the same effect on the steering. A non-rolling tyre is no longer able to have an impact on the course of the vehicle, hence ABS, with its constant release and reapplication of the brakes, allows the wheel to retain some input into the vehicle direction as it rapidly alternates between a braking component and a steering component.

A driver can simulate this clever technology on cars that are not fitted with ABS by using an old technique, discussed by many and understood by few, called cadence braking. Cadence braking is inferior to braking using the threshold technique in dry conditions (see above), but it may save your life where the road surface is greasy, snowy or icy. The technique relies on releasing the wheels once they have locked to allow some steering input. The larger the steering input, the less chance you have of retaining or regaining control, so the trick is to put in the minimum steering angle required to get you out of trouble. Many drivers, especially when they are inexperienced, default to maximum braking force when they realize the vehicle isn't stopping, and turn the wheel further when the car refuses to respond to steering input. On modern ABS-equipped cars maximum pedal pressure in an emergency is probably the best thing you can do, but on cars without such a system applying the brakes and releasing them in a regular pattern will provide similar help in stopping the vehicle. Releasing the brakes allows the wheels to spin up to road speed again, so regaining some steering control, and applying the brakes until they lock gives you that optimum period of passing through the maximum braking efficiency threshold. Cadence braking, however, is really only suitable for low grip conditions as the constant application and removal of braking forces will unsettle a car driving on higher grip surfaces, which could cause a loss of control. Ideally you should save it for emergencies on low grip surfaces, but make sure you get a chance to practise it in a safe environment when you can.

Braking changes how the weight of the car is applied to the road via the wheels, in a process called weight transfer. When brakes are applied and the weight of the vehicle is thrown forwards over the front wheels, the grip on these wheels will increase due to the additional downward loading. For this reason there is a large difference in brake capability between the front and rear axles, especially on front-engined cars. As mid- or rear-engined cars maintain quite a lot of weight over the rear axles during braking, their rear brakes can be used much more effectively during heavy braking, and as such a mid- or rear-engined car will usually stop more effectively than its front-engined equivalent. This weight transfer will affect the way you apply the brakes on the road and track. You may, for example, choose to modify the way you apply the brakes through a corner to increase the weight transferred over the front axle, which on a front-wheel drive car could reduce the inherent understeer. Remember that the point in the corner where a vehicle goes from deceleration to acceleration will transfer the weight from the front to the rear of the car, changing the handling accordingly. Modulating your braking effort during gear changes can also reduce the risk of locking the wheels, especially on a rear-wheel drive car, as reducing the braking force will momentarily send some weight back over the driven rear axle. This minimizes the risk of the rear wheels snatching if a gear change is applied in a forceful fashion. Sometimes known as transmission lock, this accidental locking of the driven wheels is sometimes initiated intentionally by rally drivers to deliberately induce a slide. For us mere mortals it is probably best avoided, if only for the sake of mechanical sympathy.

'Heel and toe' is a technique in which the right foot is used to operate the throttle pedal while simultaneously operating the brake. This allows the driver to match the rpm of the engine to the transmission speed while changing down through the gears, preventing the mismatch in rpm from locking the driven wheels. If a car is being driven consistently at the limit then only a small amount of lost traction can result in understeer or oversteer, resulting in a spin or loss of control. Usually the driver would brake with the toes while rolling his or her heel over to the throttle pedal to generate a blip of acceleration when in mid-gear change. This means that the transmission does not have to spin the engine up to a higher speed as it is already close to the required speed, reducing the risk of the driven wheels locking up. This technique can be used in everyday driving to improve mechanical sympathy with the car; if carried out correctly it will reduce the stress on mechanical components in both engine and gearbox, and increase the life of the friction plates on the clutch. If carried out in race conditions the vehicle will be more stable under braking and less likely to lock the wheels.

Braking on snow is a different skill altogether, as any budding rally driver will tell you. Counter-intuitively there are times when locking the wheels is actually preferable, as it builds up a pyramid of snow in front of the wheels, assisting with bringing the vehicle to a halt. Of course the caveat that cornering ability is lost during this still applies, and the second you let off the brakes you will probably end up driving straight over the pyramid and have to build up another to help you stop. As a result some ABS-equipped cars actually stop less effectively on snow than an equivalent car without ABS.

CHAPTER SEVENTEEN

FAULT FINDING

Due to the increased complexity of modern braking systems, fault finding has either become a lot more complicated or a lot simpler, depending on your point of view. Back in the days of rod-operated brakes the lack of complexity and the visibility of all the moving parts made diagnosis simple. A given movement at the pedal could be traced all the way through the system, with additional free play easily spotted and rectified.

Multiple adjustments were possible, allowing fine tuning. The downside, of course, was that this required frequent adjustments and balancing the system was difficult. As cars moved to hydraulic systems, it became less obvious where faults lay unless fluid was visibly leaking from the system. The latest electronically controlled systems strike fear into the hearts of many who see electronics as a black art. When these systems first came out it was certainly the case that only a few specialists were able to deal with them, the required tools were

The smaller cracks often seen on race cars or hard-worked road cars can be an early warning that a more serious crack is about to occur, as demonstrated here. Treat all cracks with caution. Monitor and replace before they get as bad as this.
ERIC JACOBSEN

expensive and they were often not available to the mass market. However, as with electronic engine management, aftermarket manufacturers put more effort into providing systems to read them as more designs flooded the market. With the proliferation of OBD-II and EOBD, standardized on-board diagnostics systems made mandatory in most of the world, it is now common for a garage or consumer to identify exactly which component on the vehicle is at fault without getting their hands dirty. As part of the EOBD regulations, mandatory in Europe on petrol vehicles since 2001 and diesels since 2004, a series of codes are generated when faults are downloaded from the vehicle using a special scanner (available for considerably less than £100 for a domestic version suitable for the home mechanic). The codes generated are usually of the format Pxxx for power-train and Cxxx for chassis, where x represents the number of the fault. Codes starting P00xx up to P04xx are to do with engine management and outside the scope of this book, codes from P05xx to P08xx could potentially have some relevance as they relate to power-train and speeds. Of most interest to us, however, are the codes beginning with Cxxxx. These C codes are listed fully in the Appendix. With this information the vehicle will actually self-diagnose, often saving a great deal of time and effort by pointing you in the right direction. Of course, like many systems it is not infallible: sometimes it will miss faults that a competent mechanic might pick up, other faults it can misdiagnose. Use the information that it provides with care and don't buy expensive components without further investigation of the issues it diagnoses.

Where there is no machine available, or further diagnosis is required, some understanding of common symptoms can help you work out what is wrong. If a vehicle pulls to one side under braking it is usually due to an imbalance in the brakes, assuming that the suspension bushes and such are all in good serviceable condition. The easiest thing to check is that the discs are being comprehensively swept; if the disc is shiny on one wheel but shows evidence of rusty unswept areas on the other side of the car, then obviously something is amiss. Ensure the pads are in good order, the pistons have not seized and that floating calipers are sliding properly. A new pair will be required if the disc is scored or the rust serious. Don't forget to check not only the side of the disc you can easily see, but also the side that is hidden, facing towards the centre of the car. If the disc is shiny on one side but not the other, then a seized caliper is possible, especially if it is a floating single piston design. If none of these things are present, then the issue could be one of suspension and steering geometry or wear.

If excessive pedal pressure is required to bring the vehicle to a stop, first check the servo. As mentioned in the section on vacuum servos (see Chapter 12), it is usually possible to check its operation by pressing the pedal repeatedly while the engine is not running, until the pedal is more difficult to move. Maintaining a firm pedal pressure to start the car, the pedal should sink slightly to the floor once the engine is running. If the pedal does not sink to the floor the servo or its associated vacuum pipework could be at fault. Listen for whistling noises around the pipework and from the servo itself, indicating an air leak. Serious air leaks from the servo could actually result in the vehicle running rough as unmetered air enters the induction system. On high powered vehicles, specifically those that are turbocharged or that run large throttle openings for any length of time, it could be that the servo is not able to retain a sufficient vacuum depression to operate. The servo builds up this vacuum when the throttle is fully or nearly closed, creating a pressure in the inlet manifold that is lower than the ambient air. At wide open throttle this vacuum is not present, and therefore servo assistance may be compromised. This is not normally an issue, as most don't want to brake and operate full throttle at the same time, but if a greater reserve of vacuum depression is required then a vacuum reservoir with a non-return valve should be used. This is a sturdy pressure vessel that stores the vacuum

depression, when possible, for use at any time by the servo. Servos tend to store a certain amount of vacuum themselves, hence the usual requirement for a non-return valve in the vacuum line.

Another reason for increased pedal effort can occur on a split braking system. Many braking systems today use a diagonal split system, so that in the event of one system failing the other system, which is unconnected to the first, can still operate. Each system has a single front and rear brake attached, diagonally split across the car. If the front left-hand/rear right-hand circuit fails, the front right-hand/rear left-hand circuit will still operate, allowing the car to be brought to a balanced stop without excessive braking instability. The downside of this system is that some drivers who are not very attuned to the performance of their car may continue to use the vehicle without realizing there is an issue. Usually, however, the significant increase in braking effort required is enough to tell people there is a fault. Some vehicles use pressure sensors in each line that remain balanced when both lines are operating correctly. In the event of one line failing, the pressure sensor is deflected over to the faulty side, illuminating a warning lamp in the dash.

Contamination of the pad or disc, perhaps from oil or lubricants used while servicing or cleaning the vehicle, can cause increased pedal effort. If the vehicle has just been washed, the brakes may need to be lightly applied a number of times to dry the discs. Some high-end manufacturers, such as BMW, use a system that can sense moisture on the brakes and applies modulating pressure to lightly, and usually imperceptibly, apply the pads to the disc. This dries the brakes without any input from the driver, allowing them to stop safely next time they need to do so.

Excessive pedal pressure can be a result of the wrong grade of friction material being used. Some race and track pad materials require a high level of heat before they will retard the vehicle effectively, but they tend to be very aggressive once hot, and resistant to fade. Ensure you are using the correct pad material to suit the day-to-day application. Having race spec pads might be good for bragging, but it isn't so great on the daily commute.

APPENDIX: EOBD/OBD-II CODES (BRAKING SYSTEM RELATED)

C0000: Vehicle Speed Information Circuit Malfunction

C0035: Left Front Wheel Speed Circuit Malfunction

C0040: Right Front Wheel Speed Circuit Malfunction

C0041: Right Front Wheel Circuit Range out of limits and/or Performance degradation

C0045: Left Rear Wheel Speed Circuit Malfunction

C0046: Left Rear Wheel Speed Sensor Circuit Range out of limits and/or Performance degradation

C0050: Right Rear Wheel Speed Circuit Malfunction

C0051: Left Front Wheel Speed Sensor Circuit Range out of limits and/or Performance degradation

C0060: Left Front Abs Solenoid #1 Circuit Malfunction

C0065: Left Front Abs Solenoid #2 Circuit Malfunction

C0070: Right Front Abs Solenoid #1 Circuit Malfunction

C0075: Right Front Abs Solenoid #2 Circuit Malfunction

C0080: Left Rear Abs Solenoid #1 Circuit Malfunction

C0085: Left Rear Abs Solenoid #2 Circuit Malfunction

C0090: Right Rear Abs Solenoid #1 Circuit Malfunction

C0095: Right Rear Abs Solenoid #2 Circuit Malfunction

C0110: Pump Motor Circuit Malfunction

C0121: Valve Relay Circuit Malfunction

C0128: Low Brake Fluid Circuit Low

C0141: Left Traction Control System Solenoid #1 Circuit Malfunction

C0146: Left Traction Control System Solenoid #2 Circuit Malfunction

C0151: Right Traction Control System Solenoid #1 Circuit Malfunction

C0156: Right Traction Control System Solenoid #2 Circuit Malfunction

C0161: ABS or Traction Control Brake Switch Circuit Malfunction

C0221: Right Front Wheel Speed Sensor Circuit Open

C0222: Right Front Wheel Speed Signal Missing

C0223: Right Front Wheel Speed Signal Erratic

C0225: Left Front Wheel Speed Sensor Circuit Open

C0226: Left Front Wheel Speed Signal Missing

C0227: Left Front Wheel Speed Signal Erratic

C0229: Drop Out of Front Wheel Speed Signals

C0235: Rear Wheel Speed Signal Circuit Open

C0236: Rear Wheel Speed Signal Circuit Missing

C0237: Rear Wheel Speed Signal Erratic

C0238: Wheel Speed Mismatch

C0241: Electronic Braking Control Module Control Valve Circuit

C0245: Wheel Speed Sensor Frequency Error

C0254: Electronic Braking Control Module Control Valve Circuit

C0265: Electronic Braking Control Module Relay Circuit

C0266: Electronic Braking Control Module Relay Circuit

C0267: Pump Motor Circuit Open/Shorted

C0268: Pump Motor Circuit Open/Shorted

C0269: Excessive Dump/Isolation Time

C0271: Electronic Braking Control Module Malfunction

C0272: Electronic Braking Control Module Malfunction

C0273: Electronic Braking Control Module Malfunction

C0274: Excessive Dump/Isolation Time

C0279: Power-train Configuration Not Valid

C0281: Brake Switch Circuit

C0283: Traction Switch Shorted to Ground

C0284: Electronic Braking Control Module Malfunction

C0286: ABS Indicator Lamp Circuit Shorted High

C0287: Delivered Torque Circuit

C0288: Brake Warning Lamp Circuit Shorted High

C0290: Lost Communications

C0292: Lost Communications

C0291: Lost Communications

C0297: Power-train Configuration Data Not Received

C0298: Power-train Indicated Traction Control Malfunction

C0300: Rear Speed Sensor Malfunction

C0305: Front Speed Sensor Malfunction

C0306: Motor A or B Circuit

C0308: Motor A/B Circuit Low

C0309: Motor A/B Circuit High

C0310: Motor A/B Circuit Open

C0315: Motor Ground Circuit Open

C0321: Transfer Case Lock Circuit

C0323: T-Case Lock Circuit Low

C0324: T-Case Lock Circuit High

C0327: Encoder Circuit Malfunction

C0357: Park Switch Circuit High

C0359: Four Wheel Drive Low Range Discrete Output Circuit

C0362: Four Wheel Drive Low Range Discrete Output Circuit High

C0367: Front Axle Control Circuit High

C0374: General System Malfunction

C0376: Front/Rear Shaft Speed Mismatch

C0379: Front Axle System

C0387: Unable To Perform Shift

C0472: Steering Speed Sensor Signal Low

C0473: Steering Speed Sensor Signal High

C0495: EVO Tracking Error

C0498: Steering Assist Control Actuator Feed Circuit Low

C0499: Steering Assist Control Solenoid Feed Circuit High

C0503: Steering Assist Control Solenoid Return Circuit Low

C0504: Steering Assist Control Solenoid Return Circuit High

C0550: ECU Malfunction: Internal Write/Checksum Malfunction

C0559: EEPROM Checksum Error

C0563: Calibration Rom Checksum Error

C0577: Left Front Solenoid Circuit Low

C0578: Left Front Solenoid Circuit High

C0579: Left Front Solenoid Circuit Open

C0582: Right Front Solenoid Circuit Low

C0583: Right Front Solenoid Circuit High

C0584: Right Front Solenoid Circuit Open

C0587: Left Rear Solenoid Circuit Low

C0588: Left Rear Solenoid Circuit High

C0589: Left Rear Solenoid Circuit Open

C0592: Right Rear Solenoid Circuit Low

C0593: Right Rear Solenoid Circuit High

C0594: Right Rear Solenoid Circuit Open

C0611: Vin Information Error

C0615: Left Front Position Sensor Malfunction

C0620: Right Front Position Sensor Malfunction

C0625: Left Rear Position Sensor Malfunction

C0628: Level Control Position Sensor Circuit High

C0630: Right Rear Position Sensor Malfunction

C0635: Left Front Normal Force Circuit Malfunction

C0638: Left Front Normal Force Circuit High

C0640: Right Front Normal Force Circuit Malfunction

C0643: Right Front Normal Force Circuit High

C0655: Level Control Compressor Relay Malfunction

C0657: Level Control Compressor Circuit Low

C0658: Level Control Compressor Circuit High

C0660: Level Control Exhaust Valve Circuit Malfunction

C0662: Level Control Exhaust Valve Circuit Low

C0663: Level Control Exhaust Valve Circuit High

C0665: Chassis Pitch Signal Circuit

C0690: Damper Control Relay Circuit Malfunction

C0691: Damper Control Relay Circuit Range

C0693: Damper Control Relay Circuit High

C0695: Position Sensor Over Current (8 volt Supply)

C0696: Position Sensor Over Current (5 volt Supply)

C0710: Steering Position Signal Malfunction

C0750: Tyre Pressure Monitor Sensor Failure

C0755: Tyre Pressure Monitor Sensor Failure

C0760: Tyre Pressure Monitor Sensor Failure

C0765: Tyre Pressure Monitor Sensor Failure

C0800: Device Power #1 Circuit Malfunction

C0896: Electronic Suspension Control Voltage Is Outside Normal Range

C1211: ABS Indicator Lamp Circuit Malfunction

C1214: System Relay Contact or Coil Circuit Open

C1217: Pump Motor Shorted to Ground

C1218: Pump Motor Circuit Shorted to Voltage or Motor Ground Open

C1221: Left Front Wheel Speed Sensor No Input Signal

C1222: Right Front Wheel Speed Sensor No Input Signal

C1223: Left Rear Wheel Speed Sensor No Input Signal

C1224: Right Rear Wheel Speed Sensor No Input Signal

C1225: Left Front Excessive Wheel Speed Variation

C1226: Right Front Excessive Wheel Speed Variation

C1227: Left Rear Excessive Wheel Speed Variation

C1228: Right Rear Excessive Wheel Speed Variation

C1232: Left Front Wheel Speed Open/Short Circuit

C1233: Right Front Wheel Speed Open/Short Circuit

C1234: Left Rear Wheel Speed Open/Short Circuit

C1235: Right Rear Wheel Speed Open/Short Circuit

C1236: Low System Supply Voltage

C1237: High System Supply Voltage

C1238: Brake Thermal Model Exceeded

C1241: Variable Effort Steering Circuit Malfunction

C1242: Pump Motor Circuit Open

C1243: Brake Pressure Modulator Valve Pump Motor Stalled

C1244: Power-train Indicated Engine Drag Control Malfunction

C1246: Brake Lining Wear Circuit Open

C1248: Electronic Brake Control Module Illuminated Brake Warning

C1251: RSS Indicated Malfunction

C1252: Left Front Normal Force Malfunction

C1253: Right Front Normal Force Malfunction

C1254: Abnormal Shutdown Detected

C1255: Electronic Braking Control Module Internal Malfunction

C1256: Electronic Braking Control Module Internal Malfunction

C1261: Left Front Inlet Valve Solenoid Malfunction

C1262: Left Front Outlet Valve Solenoid Malfunction

C1263: Right Front Inlet Valve Solenoid Malfunction

C1264: Right Front Outlet Valve Solenoid Malfunction

C1265: Left Rear Inlet Valve Solenoid Malfunction

C1266: Left Rear Outlet Valve Solenoid Malfunction

C1267: Right Rear Inlet Valve Solenoid Malfunction

C1268: Right Rear Outlet Valve Solenoid Malfunction

C1271: Left Front Traction Control System Master Cylinder Isolation Valve Malfunction

C1272: Left Front Traction Control System Prime Valve Malfunction

C1273: Right Front Traction Control System Master Cylinder Isolation Valve Malfunction

C1274: Right Front Traction Control System Prime Valve Malfunction

C1276: Delivered Torque Signal Circuit Malfunction

C1277: Requested Torque Signal Circuit Malfunction

C1278: Traction Control System Temporarily Inhibited by PCM

C1281: VSES Sensors Uncorrelated

C1282: Yaw Rate Sensor Bias Circuit Malfunction

C1283: Excessive Time to Centre Steering

C1284: Lateral Accelerometer Sensor Bias Malfunction

C1285: Lateral Accelerometer Sensor Circuit Malfunction

C1286: Steering Sensor Bias Malfunction

C1287: Steering Sensor Rate Malfunction

C1288: Steering Sensor Circuit Malfunction

C1291: Open Brake Lamp Switch Contacts During Deceleration

C1292: Brake Fluid Pressure Sensor Circuit

C1293: Code C1291 Set in Previous Ignition Cycle

C1294: Brake Lamp Switch Circuit Always Active

C1295: Brake Lamp Switch Circuit Open

C1296: Brake Fluid Pressure Sensor Circuit

C1297: PCM Indicated Brake Extended Travel Switch Failure

C1298: PCM Class 2 Serial Data Link Malfunction

C1326: Battery Out Of Range

C1650: RSS Control Module Fault

C1658: RSS Control Module Calibration Fault

C1710: Left Front Damper Actuator Short Circuit to Live

C1711: Left Front Damper Actuator Short Circuit to Chassis

C1712: Left Front Damper Actuator Open Circuit

C1715: Right Front Damper Actuator Short Circuit to Live

C1716: Right Front Damper Actuator Short Circuit to Chassis

C1717: Right Front Damper Actuator Open Circuit

C1720: Left Rear Damper Actuator Short Circuit to Live

C1721: Left Rear Damper Actuator Short Circuit to Chassis

C1722: Left Rear Damper Actuator Open Circuit

C1725: Right Rear Damper Actuator Short Circuit to Live

C1726: Right Rear Damper Actuator Short Circuit to Chassis

C1727: Right Rear Damper Actuator Open Circuit

C1735: Compressor Relay Short to Battery

C1736: Compressor Relay Short to Chassis/Open Circuit

C1737: Exhaust Solenoid Valve Short to Live

C1738: Exhaust Solenoid Valve Short to Chassis

C1743: Speed Signal Fault

C1744: Lift/Dive Signal Discrete Fault

C1760: Left Front Position Sensor Input Fault

C1761: Right Front Position Sensor Input Fault

C1762: Left Rear Position Sensor Input Fault

C1763: Right Rear Position Sensor Input Fault

C1768: Position Sensor Supply Fault

C1780: Loss of Steering Position Signal

C1782: Iccs2 DI Left Output Short Circuit to Live

C1783: Iccs2 DI Left Output Short Circuit to Earth

C1784: Iccs2 DI Right Output Short Circuit to Live

C1785: Iccs2 DI Right Output Short Circuit to Earth

C1786: Damper Control Relay Fault

C1787: Damper Control Relay Short to Earth

C1788: Damper Control Relay Short to Live

INDEX